A Builder's Guide to
Residential HVAC Systems

NAHB Research Center

A Service of

HOME BUILDER PRESS

NAHB

Home Builder Press®
National Association of Home Builders
1201 15th Street, NW
Washington, DC 20005-2800
(800) 223-2665

...ide accurate ...ard to the subject matter covered. It is sold with the understanding that the publisher is not engaged in rendering legal, accounting, or other professional service. If legal advice or other expert assistance is required the services of a competent professional person should be sought.

—From a Declaration of Principles jointly adopted by a Committee of the American Bar Association and a Committee of Publishers and Associations.

A Builder's Guide to Residential HVAC Systems
ISBN 0-86718-423-X

Library of Congress Cataloging-in-Publication Data

A builder's guide to residential HVAC systems / NAHB Research Center.
        p.   cm.
    Includes bibliographical references.
    ISBN 0-86718-423-X (alk. paper)
    1. Dwelling—Heating and ventilation.
2. Dwelling—Air conditioning. I. NAHB Research Center.
    TH7684.D9B85 1997
    697—dc21   96-36821
                                        CIP

For further information please contact:
 NAHB Research Center, Inc.
 (800) 638-8556
   or
 Home Builder Press®
 National Association of Home Builders
 1201 15th Street, NW
 Washington, DC 20005-2800
 (800) 223-2665

1/97 Edington-Rand/McNaughton/1,500

NAHB Research Center web site:
 http://www.nahbrc.com

Home Builder Bookstore web site:
 http://www.nahb.com/builderbooks

Additional copies of this publication are available from either the NAHB Research Center or NAHB/Home Builder Press. NAHB members receive a 20 percent member discount on publications purchased through Home Builder Press. Quantity discounts also are available through Home Builder Press. For more information about discounts, please write to the Director of Marketing, Home Builder Press, 1201 15th Street, NW, Washington, DC 20005-2800, or call 1-800-368-5242, ext. 394. When ordering or reordering this publication, please provide the following information: title, price, quantity, NAHB membership number (as applicable), and complete mailing address, including zip code.

Contents

Figures

Acknowledgments

This guide was prepared by the NAHB Research Center, Inc., for publication by the National Association of Home Builders. The Research Center wishes to acknowledge the assistance and technical information obtained from numerous people in the home-building and heating, ventilating, and air-conditioning industries. We also wish to acknowledge the suggestions and ideas received from other trade associations, various utility companies and their associations, and a number of manufacturers of HVAC-related products. Reviewers included Russell Arkin, Arkin Homes; Joseph Chudnow, Chudnow Construction Corporation; Robert Hammon, ConSol; William Hauke, Jr., Building Supply, Inc.; Harold W. Heiss, American Electric Power; Ted Hicks, NAHB Research Center; Thomas Kenney, NAHB Research Center; Jim Leach, Wonderland Homes; Mark Modera, Lawrence Berkeley National Laboratory; and Loren Swanson, Southern Michigan Company. While acknowledging the advice and suggestions of these people and organizations, we wish to emphasize that this in no way implies their endorsement, since the concepts and statements presented herein are those of the Research Center and authors Christine Barbour, Dan Cautley, Jay Crandell, James Lyons, and Matt Pesce.

A Builder's Guide to Residential HVAC Systems was produced under the general direction of Kent Colton, NAHB Executive Vice President/CEO and Liza Bowles, President, NAHB Research Center, Inc., in association with staff members Jim DeLizia, Staff Vice President, NAHB Member and Association Relations; Adrienne Ash, Assistant Staff Vice President, NAHB Publishing and Information Services; Rosanne O'Connor, Director of Publications, NAHB Home Builder Press; Jay Crandell, Project Manager, NAHB Research Center; Matthew M. Pesce, Project Engineer, NAHB Research Center; Sharon Lamberton, Assistant Director of Publications and Project Editor, NAHB Home Builder Press; David Rhodes, Art Director, NAHB Home Builder Press; and Carolyn Kamara, Editorial Assistant, NAHB Home Builder Press.

Introduction

The purpose of this guide is to provide home builders with sufficient information to enable them to evaluate Heating Ventilating and Air Conditioning (HVAC) system designs and to work more effectively with HVAC contractors. Tables, charts, guidelines, and examples of duct layouts are presented throughout this text. Ductwork and alternative duct systems are emphasized, but general information is also provided on equipment selection, installation costs, operating costs, and the design of the total HVAC system. New information presented in this edition details energy-efficient options related to duct systems and management issues that directly affect efficient duct design and installation. Publications on thermal insulation, designing and building energy-conserving homes, load calculations, and detailed design of duct systems are referenced for additional guidance.

This guide is not intended as a design manual for use by professional engineers or HVAC contractors. Sufficient information is presented, however, to enable builders to do a preliminary design for the system and to work proactively with HVAC contractors. The primary usefulness of this guide will be in evaluating various design options and making modifications to provide a cost-effective duct system for comfort and energy efficiency in modern homes.

Chapter 1 discusses the "nuts and bolts" of ductwork and fittings and introduces terminology unique to the ductwork trade. Relative costs, advantages, and disadvantages of available duct materials are presented. Various duct fittings and components are illustrated, with accompanying descriptions of their application and relationship to the total system.

Chapter 2 describes various design approaches to air distribution systems for residences. In addition to general guidelines and limitations affecting their design, characteristics of each type of system are discussed, including relative ease of fabrication and installation, performance, adaptability to different constructions, concealment characteristics, and relative cost. This chapter also has a section on return-air systems, including a comparison guide on the advantages and disadvantages of central and individual returns, the use of structural spaces for return-air passages in place of ductwork, and design tips for improving sound attenuation.

Chapter 3 provides brief discussions on heating and cooling loads, analyses of typical and alternative mechanical equipment, and final system design and sizing. Different equipment types and configurations are discussed and a description of air patterns generated by supply outlets and returns, guidelines for register and grille locations, and tips on duct sizing are included.

Chapter 4 discusses system efficiency improvements for both ductwork and equipment. Several ideas are presented, including accurately sized equipment and measures to improve duct performance. Testing techniques such as blower door tests and duct blaster tests are discussed relative to their usefulness in optimizing HVAC system design.

Chapter 5 presents the management factors affecting duct system quality. Issues such as contractor selection, contracts, planning, design, coordination, and inspection are discussed. Management tips and tools are given for adoption and modification by builders.

Chapter 6 discusses the marketing of system efficiency. Issues include how customers view energy efficiency, marketing strategies, communicating features and benefits, utility demand-side management programs, home energy ratings, energy financing, and how to develop a complete energy efficiency marketing package.

Chapter 1

Components of an Air Distribution System

An air distribution system consists of supply and return plenums, ductwork and fittings, boots, grilles, and registers. Supply ducts deliver air to the spaces that are to be conditioned. Return ducts remove room air and move it back to the equipment for filtering and reconditioning. Registers are positioned to mix conditioned air with room air. Basic duct materials and duct system components are discussed in this section.

DUCT MATERIALS

Selection of duct material is based on its price, performance, and installation requirements. Commonly available duct materials are sheet metal, fiberglass duct board, flexible duct, and duct insulation. Materials used for wall and building cavity returns are not discussed.

Sheet Metal

The most common duct material is cold-formed, galvanized sheet steel. Sheet metal trunks and run-outs are typically 4- or 8-feet long. Rectangular ducts are usually connected using drive- and "S"-slip joints, while round ducts are connected with crimp joints. Seams for both rectangular and round ducts are typically pocket locks, as shown in Figure 1.1. Trunk take-off or run-out connections are either flap joints or dove-tailed joints (not shown).

Sheet metal ducts have several advantages because they are:

- Made of relatively inexpensive materials.
- Rugged and durable.
- Widely available in many sizes and gauges.
- Easily fabricated into both rectangular and round shapes.
- Readily formed into special transitional fittings.
- Can pass through fire barriers.

    In addition, their smooth surface offers low resistance to air flow.

Some disadvantages of sheet metal ducts are:

- Each connection, joint, and seam has potential leakage.

1

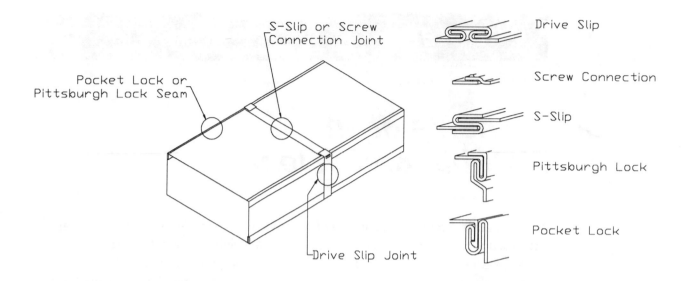

Figure 1.1: Sheet Metal Duct Connections

- Systems must be field- or shop-insulated when placed in attics, crawlspaces, and unconditioned spaces.
- There is higher sound transmission.
- They are less flexible in routing and have higher installation costs compared to some systems.
- During comfort cooling, uninsulated metal ducts are prone to developing condensation on the outside of the duct.

Sheet metal is standard in fitting and ductwork sizing guides. When installing alternate materials, allowance sometimes must be made for higher internal friction or increased friction loss because of material surface characteristics. Metal ductwork is adaptable to most HVAC systems used in residential construction. It is particularly economical and reasonably energy-efficient when installed in conditioned spaces, where the added cost of insulation or sealants is not required. When located in unconditioned spaces, metal ducts should be sealed and must be insulated. HVAC contractors may use fiberglass insulating jackets or liners for this purpose. Externally applied insulation usually has a vapor barrier. Metal ducts can also be embedded in concrete slabs, but if this is done groundwater protection is necessary and the ducts must be securely tied down to prevent floating when concrete is poured.

The trade association for union sheet metal duct contractors is the Sheet Metal and Air Conditioning Contractors National Association (SMACNA):

Sheet Metal and Air Conditioning Contractors National Association
4201 Lafayette Center Drive
Chantilly, VA 22021
(703) 803-2980
http://www.smacna.org

SMACNA offers documents for sheet metal duct design, installation, and fabrication. Another industry trade organization, the Air Distribution Institute (ADI), represents some of the larger sheet metal manufacturers. The Air Distribution Institute can be contacted at the following location:

Air Distribution Institute
4415 West Harrison Street, Suite 242C
Hillside, IL 60162
(708) 449-2933

Fiberglass Duct Board

Fiberglass duct board is insulated and sealed as part of its construction. Trunks and run-outs are available in 4- and 8-foot sections. Rectangular ducts are usually connected using shiplap joints and pressure-activated tapes (see Figure 1.2). Corner joints (seams) may be shiplapped or V-grooved. The seams are then taped. Closure systems are tested and stamped under Underwriter's Laboratory Standard 181 A/P. Each of the above-mentioned joints is described more fully in North American Insulation Manufacturers Association's (NAIMA) *Fibrous Glass Residential Duct Construction Standards*.

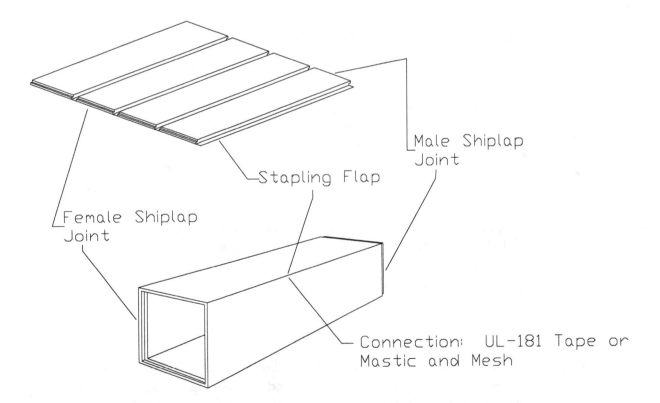

Figure 1.2: Fiberglass Duct Board Connections

Rigid fiberglass duct board is commonly used to form rectangular trunk duct and fittings such as register boots and intermediate distribution boxes for insulated flex duct branches located in unconditioned spaces such as attics and crawlspaces. The most common stock size is 1-inch thick, which provides a nominal R-value of 4.3. Duct board of 1½ inches (R-6.5) is also available. The necessary cuts, laps, and grooves can be made with a few simple hand tools. Special production shop equipment is available. Joints are generally secured by stapling and taping, using reinforced vapor-barrier tape similar to the outer skin material. Rigid round fiberglass duct is factory-formed from materials similar to duct board and is typically used for individual branch ducts in extended plenum duct board systems. Cutting and joining techniques are similar to those used for duct board.

For installations that require duct insulation, such as attic or crawlspace distribution systems, fiberglass ducts are cost-effective. Material costs are higher than those for sheet metal, but installed costs may be comparable, especially when sheet metal must be insulated.

Some advantages of duct board include:

- It is lightweight and particularly adaptable to attic systems.
- A vapor barrier is included as an integral part of the duct material.
- It provides excellent sound attenuation.

Some of its disadvantages are:

- Its durability and longevity are highly dependent on its closure system (tapes and mastics).
- It may be damaged or crushed during construction.

The primary trade association involved with fiberglass ductwork is the North American Insulation Manufacturers Association (NAIMA). NAIMA offers information on the design and installation of fibrous glass residential duct systems. SMACNA also offers information on the design and installation of duct board systems. NAIMA may be contacted at the following address:

North American Insulation Manufacturers Association
44 Canal Center Plaza, Suite 310
Alexandria, VA 22314
(703) 684-0084
http://www.naima.org

Flexible Duct

Flexible duct is constructed from blanket insulation covered with a flexible vapor-barrier jacket on the outside and supported on the inside by a helix wire coil covered by a polyester or chlorinated polyethylene core. A metal collar connects flexible duct to sheet metal or duct board trunk duct, as shown in Figure 1.3. Flexible duct cannot be passed through fire barriers.

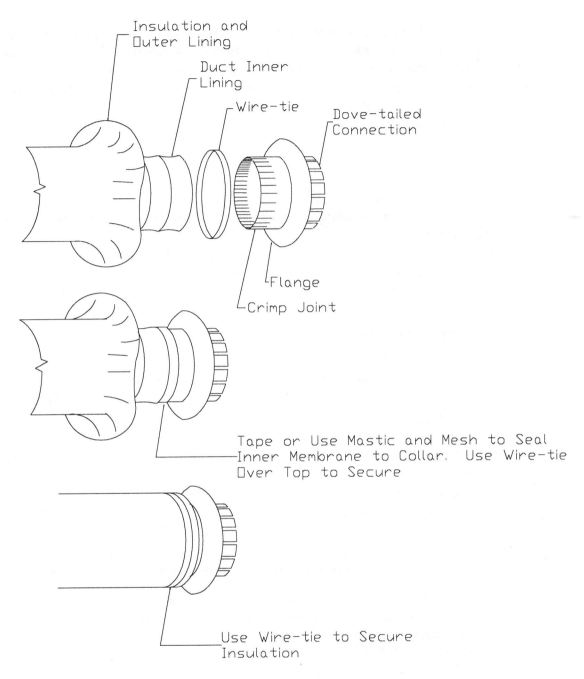

Figure 1.3: Flexible Duct Connections

Flexible duct is available in 25-foot standard length with plain, unfitted ends. Insulated duct sizes vary from 4- to 20-inch inside diameter. Insulation is available with typical R-values of 4.2 and 6.0. Uninsulated flexible duct may be used for ducts within conditioned spaces and for appliance and fan exhaust vents. It is available in typical inside diameters of 3 to 14 inches; some stronger products span the range from 2 to 20 inches.

Manufacturers provide performance curves for flow and pressure drop of straight runs of flexible duct, but little performance data is available for ducts in 45-degree elbows, 90-degree elbows, 180-degree offset elbows, or other configurations found in many actual installations.

Many installations can be simplified by using flexible duct instead of rigid ductwork. It offers the following advantages:

- There are fewer duct connections and joints.
- It is factory insulated.
- It has a low installation cost.
- It has a low material cost.
- It is suited for attic installations.

Disadvantages of flexible ductwork include:

- It is easily torn, crushed, pinched, or damaged.
- Damage to the inner lining is not visible.
- It has higher resistance to air flow and must be properly specified.
- Little pressure-drop performance data is available.

The primary trade association for flexible ductwork manufacturers is the Air Diffusion Council (ADC). ADC offers information on the design and installation of flexible air ducts. SMACNA also offers information on designing and installing flexible duct systems. The Air Diffusion Council may be contacted at the following location:

Air Diffusion Council
11 South LaSalle Street, Suite 1400
Chicago, IL 60603
(312) 201-0101

Duct Insulation

Duct insulation is available as duct liner and as exterior duct wrap. Both liner and wrap are used in residential construction. Liner is more often used to line return register boxes and awkwardly placed supply trunks. Duct wrap tends to be used on uninsulated duct materials such as sheet metal placed in attics or crawlspaces.

Fiberglass Duct Liner

Available in ½-, 1-, 1½-, and 2-inch thicknesses, duct liner is a specially treated fiberglass insulation used to line the inside of rectangular metal ductwork. Typical densities are 1.5 pcf (R-3.6/inch), 2.0 pcf (R-3.7/inch), and 3.0 pcf (R-4.2/inch).

HVAC contractors primarily use 1-inch thick fiberglass duct liners for thermal protection of ductwork that passes through unconditioned spaces. Liners that are ½-inch thick frequently are used as acoustical insulation for reducing the transmission of air and equipment noise and to line return register boxes. Duct liner is installed by applying a special adhesive to the inside of ductwork or to flat stock prior to forming. The adhesive may be supplemented with metal fasteners at critical points.

Fiberglass Duct Wrap

Duct wrap is available in 1½-, 2-, 2¼-, and 3-inch rolls. Each roll is 4 feet wide. Typical insulation densities are 0.75 pcf (R-3.6/inch), 1.0 pcf (R-3.8/inch), and 1.5 pcf (R-4.1/inch).

Duct wrap is a fiberglass blanket insulation which comes both unfaced and faced. Faced insulation may be covered with either a foil-scrim-Kraft (FSK) or vinyl-vapor barrier. It is used primarily to insulate metal ductwork that is located in unconditioned spaces. Duct wrap provides better thermal protection than duct liner but is of little benefit acoustically. It is installed by wrapping the outside of the ductwork and taping the joints.

Ductwork in Concrete Slabs

Plastic and plastic-coated metal ducts are used where ducts are to be embedded in a concrete slab. These materials offer the following advantages:

- Specialty products are available.
- Joints are readily waterproofed to prevent the infiltration of groundwater.

Disadvantages of these materials include:

- Ducts must be run on a slope to allow condensate drainage.
- They are susceptible to high groundwater.
- They are difficult to repair or replace.
- HVAC work must be coordinated during foundation construction.

MAJOR DUCTWORK COMPONENTS

Duct systems generally include supply and return plenums, trunk ducts, branch ducts, register boots, registers, grilles, and various fittings that join segments together. Figure 1.4 shows a composite supply duct system that incorporates a variety of fittings used in residential work.

Where changes in direction or reductions occur in ductwork, a transition fitting is desirable to minimize air friction and turbulence within the duct system. Individual components of the duct system are discussed in the following sections on ductwork, fittings, and registers. Detailed illustrations may be found in catalogs of various product manufacturers.

Ductwork

Plenum ducts, trunk ducts, and branch ducts are typically used in residential construction.

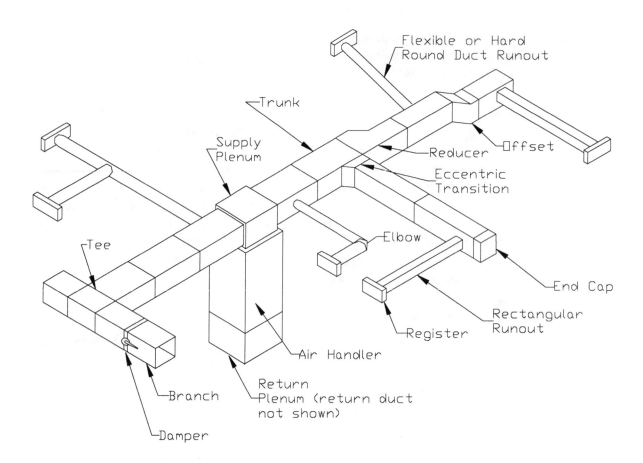

Figure 1.4: Major Ductwork Components

Plenum Ducts

A plenum is a collector box attached to the supply or return side of the air handler. It distributes or collects air from major trunk ducts. A plenum is used on the supply side of most systems, and in certain instances on the return side. It must be custom fabricated to fit the opening size of the equipment cabinet and may be insulated, depending on its location in the building. When air conditioning is available, the evaporator coil is usually housed in the plenum. Plenum ducts are usually rectangular.

Trunk Ducts

Trunk ducts are the main supply (or return) ducts that connect directly to the plenum and from which branch ducts extend to individual outlets. Trunk ducts and fittings are normally rectangular to provide for:

- Ease of fabrication, handling, and installation.
- Concealment within structural spaces.
- A less cumbersome, neat-looking installation.

Trunk duct is available in a wide range of sizes. Residential trunk duct is fabricated in sections starting at 8 inches high and 8 inches wide, with larger sizes available in increments of 2 inches. Each section is normally 4- or 8-feet long, depending on the fabricator's stock sizes. Metal trunk duct is often fabricated in two L-shaped halves to simplify handling and minimize storage space, with final assembly being made in the field. Depending on metal gauge, metal ducts 12 inches and wider should have a cross break (an X crease on each face) to add rigidity and eliminate "oil canning" noises when the blower starts and stops. Trunk ducts as well as plenums may be fabricated from fiberglass duct board or flex duct.

Branch Ducts

Branch ducts are the smaller, individual ducts that run from the main trunk duct to individual outlets. Round sheet-metal or flexible ducts are most popular for this purpose, but round fiberglass ducts also are available. Round branch ducts have some important advantages, including:

- They have good air flow characteristics.
- They require fewer types of fittings.

HVAC contractors often use oval or thin rectangular ducts as interior wall stacks. These ducts have higher pressure losses than round or square ducts but fit easily within a 2x4 stud space.

Fittings

Fittings used in residential construction include starting collars, flexible connectors, dampers, elbows, reducers, end caps, takeoff fittings, stackheads, and register boots.

Starting Collar

This fitting is attached to a rectangular hole cut into the side of the plenum and provides a transition to match the size of the trunk duct being installed. It is sometimes omitted to reduce cost, with the trunk duct connecting directly into the plenum, but this increases resistance to air flow.

Flexible Connector

This fitting consists of a canvas material bonded to a metal fitting at each end. An optional item, it is installed between the starting collar and trunk duct to isolate the duct system from mechanical vibration and equipment noise. It can also provide some installation tolerance where the two ends are not in exact alignment.

Damper

Dampers may be used in trunk or branch ducts where zonal or individual balancing of air flow is desired. They are best used in supply trunks where large flows may be controlled with less hardware and where any noise that is generated is less a problem. Trunk dampers are used to:

- Provide means of adjusting air volumes when changing between heating and cooling seasons, especially in multilevel homes.
- Allow for corrective balancing when trunk ducts are not ideally sized.
- Permit balancing of standardized duct systems in tract homes where house orientation or exposure affect heating or cooling requirements.

Elbows

Horizontal elbows are installed to change direction within a space such as a basement or an attic. Vertical elbows are installed where trunk ducts change elevation, as in a split-level home, or where they turn upward, as in a central return. Rectangular elbows are usually fixed, while round elbows may be adjustable. Any angle of elbow can be manufactured but 45- and 90-degree elbows are most common.

Reducers

Reducers allow for a smooth transition from one trunk duct size to a smaller size. They are fabricated for a specific duct size at each end. Reducing adapters are more universal and may be used to reduce any duct width by a specified amount, typically 4 inches, without requiring a special fitting.

End Caps

End caps are used at the end of a trunk and are usually field-fabricated.

Takeoff Fittings

Takeoff fittings are used to tap into the side or top of a trunk duct. They form a transition from a rectangular trunk duct to a branch duct. Top takeoffs are used where branch ducts are to be located within a joist space above the trunk duct; side takeoffs are used where this is not possible or not required. To cut costs, installers may sometimes eliminate the takeoff fitting and connect the round duct or elbow directly into the trunk duct. However, this practice can result in increased air leakage and poor air flow.

Stackheads

Stackheads are designed to terminate branch ducts where outlets occur in the wall. A stackhead has metal ears for nailing to studs and attaching register screws. Stackheads can be fabricated for attachment to thin rectangular or oval ducts.

Register Boots

Register boots provide a transition from round or rectangular duct to a rectangular opening for a floor or ceiling register. Some boots incorporate an integral balancing damper. However, where accessible, a balancing damper located in the branch duct close to the trunk provides more effective and quieter air volume control.

Outlet and Return Grilles, Registers, and Diffusers

Grilles, registers, and diffusers are louvered metal units used at supply outlets and return inlets. Grilles are generally used to cover return inlets, while registers or diffusers are commonly used at supply outlets to control air delivery. A wide choice of grilles and register configurations and sizes is available for residential use. Generally constructed from lighter-gauge materials, residential grilles and registers offer fewer adjustment features than more costly commercial units. Simple painted steel units are most common, but decorative units also are available in other materials. Decorative units generally have reduced air flow as compared to more standard designs.

Return Grilles

A grille is an inlet cover with fixed louvers and no damper mechanism. Grilles are normally used at return air intakes to conceal the duct opening. Grilles are sometimes used at supply outlets but provide no means of regulating air flow. Grilles are available in a wide variety of sizes, including typical register sizes, and are available for wall, ceiling, and floor applications. Some return grilles may include filter attachments. These units, which combine a hinged return air face with a filter rack, permit servicing the HVAC system filter at the return grille. They are ideal for installations where the mechanical equipment is in locations such as an attic or a crawlspace.

Supply Registers

A register is a grille with an operable damper. The air delivery pattern from a register can range from uniform to fan-shaped, depending on louver configuration. Two types of dampers are commonly used on registers: single-blade dampers and opposed-blade dampers. Opposed-blade dampers provide more uniform air flow. Registers are available for floor, ceiling, baseboard, and wall applications.

Diffusers

Diffusers are a special type of register that delivers air parallel to adjacent surfaces; they are commonly used in ceiling applications. Ceiling diffusers provide superior air distribution patterns for cooling. Adjustable models are also suitable for some heating conditions. The terms *register* and *diffuser* are often interchanged. Standard ceiling diffusers are available in both round and rectangular configurations and are installed mainly in cooling installations where heating is less critical. Wide spacing of deflection louvers provides maximum free area for airflow, and directs air in a flat blanketing pattern. Curved blade ceiling diffusers have curved louvers that can be individually adjusted. This style is excellent for cooling with the blades adjusted outward. With the blades adjusted downward, this style also provides better air movement for heating. Models are available with one-, two-, three-, or four-way air-throw patterns.

Chapter 2

Selecting the Right System

This chapter describes the basic types of duct systems so that, as a builder, you may knowledgeably consider alternative systems with respect to house type and other design considerations. The final selection and design of the system should be worked out with a competent HVAC contractor.

The most common residential duct systems are extended plenum and radial systems; they are used extensively because of their versatility, performance, and economy. These systems and several others are illustrated in this chapter, showing their adaptability to different house types. While supply-air systems are emphasized, return-air systems are also discussed. The examples used in this chapter illustrate a broad variety of duct system possibilities, but do not necessarily represent the optimum solution for any particular case. Additional design information may be obtained from the Air-Conditioning Contractors of America:

> Air-Conditioning Contractors of America (ACCA)
> 1712 New Hampshire Avenue, NW
> Washington, DC 20009
> (202) 483-9370
> http://www.acca.org

EXTENDED PLENUM SYSTEMS

The most commonly used residential duct system is the extended plenum system. A relatively large main supply duct (trunk duct) is connected to the furnace supply plenum and serves as an extension to the plenum. The smaller branch ducts that deliver air to the individual outlets are connected into the trunk duct at various points. Adding trunk reducers can improve air flow through branch ducts closer to the air handler.

Characteristics of Extended Plenum Systems

Extended plenum systems have several advantages:

- **Simplicity.** Relatively long runs of one size rectangular trunk duct permit ease of fabrication and installation with a minimum number of sizes or special fittings.
- **Performance.** Balancing of air flow to rooms presents no major problem in the average-sized house, especially with centrally located equipment. Trunk reductions improve distribution of air among branch runs.

- **Adaptability.** The system readily adapts to most house types including one-story, two-story, and multi-level designs. Ideally suited to basement constructions, it can also be installed in crawlspaces, attics, and dropped ceilings.
- **Concealment.** Rectangular trunk ducts can be readily concealed in finished areas by bulkheading, locating within floor trusses or I-joists, and other means. Smaller branch ducts may be installed within joist and stud spaces.
- **Cost.** The extended plenum system is a low-cost system for basement, bi-level, and split-level construction. This system can also be cost-effective with slab-on-grade construction, where it may be installed in a dropped hall ceiling with potentially large energy-saving benefits. Trunk reductions add little or no extra cost and improve system performance.
- **Energy efficiency.** Extended plenum systems may be used to locate ducts in conditioned spaces (such as floor framing, basements, or serving branch duct risers in interior walls) for improved efficiency.

Design of Extended Plenum Systems

The principle design limitation of the extended plenum system is the length of single-size trunk duct. To maintain reasonable uniform air pressures in the air-distribution system, the length of a single-size trunk duct should be limited to about 24 feet. When this length is exceeded, pressure tends to build up toward the end of the duct, resulting in too much airflow in branches near the ends and insufficient airflow in branches closer to the equipment. In extreme cases where unreduced duct length is excessive, reduced pressures at branch duct takeoffs close to the equipment can actually cause air to be drawn into supply registers rather than being forced out.

A mixed extended and reducing plenum system is shown in Figure 2.1. In this application the equipment is centrally located, with a straight trunk duct serving one group of branch outlets and another reducing trunk duct serving a similar group of branch outlets.

Extended plenum systems with centrally located equipment can be used in homes up to approximately 50 feet long and still be within design limitations, depending on register locations in end rooms.

If this system is located in a finished basement area the trunk duct typically is installed close to the center girder, where it and the girder can be concealed in a bulkhead. If engineered floor trusses or I-joists are used, the trunk ducts may be located in the floor cavity, thus reducing finishing costs. In either case, the branch ducts feeding perimeter floor outlets are concealed within the joist spaces.

Sometimes it is impractical to locate the equipment centrally. Proximity to a flue on an end wall or other floorplan considerations may require that equipment be located at one end of the building. Such designs may require trunk ducts in excess of 24 feet. Under these conditions, a reducing plenum system would be required.

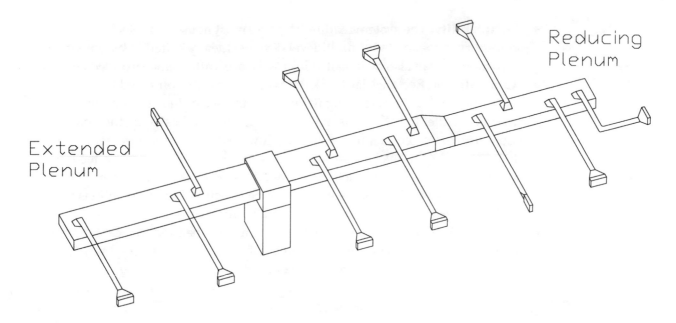

Figure 2.1: Extended and Reducing Plenum System

RADIAL SYSTEMS

Radial duct systems are the second most commonly used system in single-family homes. In radial duct systems there is no trunk duct; branch ducts that deliver conditioned air to individual outlets connect directly to the equipment plenum. Radial systems typically are used where it is not necessary to conceal ductwork, and where the equipment may be centrally located. This system is also used for in-slab constructions.

Characteristics of Radial Systems

Typical characteristics of radial systems are:

- **Simplicity.** Branch ducts are run in the most direct route from the plenum to each outlet; finished appearance of ductwork need not be considered.
- **Performance.** Air flow in the branch ducts is fairly uniform, since all branch ducts originate at a central plenum. If balancing dampers are installed in the branch ducts near the plenum, they can be serviced from a central location. Air flow is also maximized due to short, direct runs.
- **Concealment.** Radial systems typically are installed in crawlspaces and attics and below concrete slabs so no special design or finishing work is needed to conceal ductwork.
- **Adaptability.** The system is most adaptable to single-story structures with centrally located equipment. Application to other structures is limited.
- **Cost.** This is the lowest-cost system for many single-story structures. The basic simplicity of the system provides cost savings through reduced materials inventory and the use of less specialized labor.

- **Energy efficiency.** Because radial systems are typically installed in attics and crawlspaces, ducts should be well insulated and sealed to improve efficiency. When used below slabs, uninsulated ducts can be used to give a radiant heat effect, provided the slab perimeter or edge is adequately insulated.

Design of Radial Systems

Several basic design considerations affect radial duct systems:

- Equipment must be centrally located to take best advantage of the system.
- The system is most economical when applied to single-story rectangular homes.
- The system can provide economies when applied to single-level elements of two-story and split-level homes.

A typical radial system is shown in Figure 2.2. Ductwork shown in this system could be located in a crawlspace or basement or embedded in a concrete slab. Return air ducts for radial systems are typically central and are located close to the heating and cooling equipment. A variation of this system uses a perimeter loop and is suitable for below-slab installations, as shown in Figure 2.3.

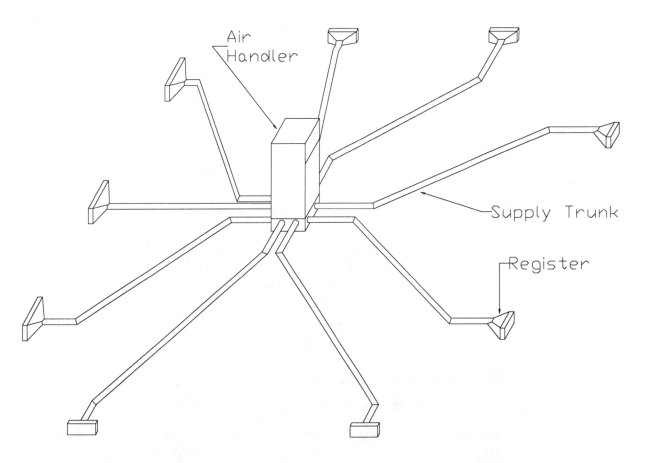

Figure 2.2: Radial System

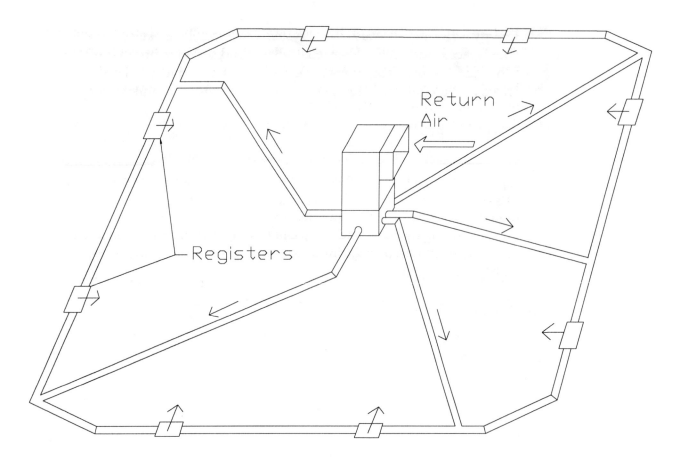

Figure 2.3: Radial Duct System With Perimeter Loop

SPIDER SYSTEMS

Spider duct systems are a variation of the radial and trunk or branch system. Spider systems use insulated flexible duct and intermediate mixing boxes. Large flex duct trunks are attached to the supply plenum and are run toward the perimeter or remote parts of the building to supply intermediate mixing boxes. Smaller branch ducts take air from the mixing boxes to the individual supply registers.

Spider systems are replacing extended plenum and radial systems in applications in attics and crawlspaces. Spider systems typically are used where it is not necessary to conceal ductwork. They may be used for equipment that has been centrally located or placed near the building perimeter. Proper design requires attention to material selection and duct design to ensure proper air flow.

Characteristics of Spider Systems

Typical characteristics of spider systems are:

* **Simplicity.** Trunks and branches may be run in the most direct route from the plenum; finished appearance need not be considered.

- **Performance.** Duct systems must be properly designed and sized to prevent restricted air flow. Ducts must be properly supported to prevent pinching of flex duct runs. Further, balancing dampers may be difficult to install.
- **Concealment.** Spider systems usually are installed in locations where concealment is not a design consideration.
- **Adaptability.** The system is most adaptable to attic and crawlspace systems where ducts can be efficiently run in the shortest available distances.
- **Cost.** This is a low-cost system for many structures. The basic simplicity of the system provides cost savings through reduced materials inventory and less labor.
- **Energy efficiency.** The system is formed from insulated flexible duct and duct board mixing boxes and is typically located in unconditioned spaces. The system only needs to be installed and properly sealed, since it comes with "built-in" insulation.

Design of Spider Systems

Several basic design considerations affect radial spider duct systems:

- Intermediate junction boxes are required; these are typically fabricated from duct board.
- Flexible duct runs should be installed in straight runs, be properly supported, and cut to accurate length.
- Kinked duct runs, extra coils, and pinched ducts increase pressure drop and reduce available system flow in improperly installed systems.
- Insulation levels for flex duct and duct board can be specified to meet energy performance requirements.

A typical spider system is shown in Figure 2.4. Ductwork shown in this system could be located in an attic or crawlspace. Return air ducts for spider systems typically are central returns using large diameter flexible duct.

RETURN AIR SYSTEMS

Proper design of the return air system is important to ensure complete system performance. Its design is stressed here because the return air system is frequently designed as an afterthought, which can result in higher operating noise levels, lower available air flow, and reduced system efficiency. Return air systems affect local air-flow patterns and should be located so as to induce a good mixing of supply air before it is returned to the equipment. Return systems must be properly sized to handle the volume of supply air.

Return air systems may be active or passive. Active return air systems are attached directly to the air handler through ductwork. Central and multiple room returns are classified as active returns. Passive return air systems are not directly attached to the air handler but aid in movement of house air toward active returns. Returns of this type include door undercuts, transfer grilles, and *jump ducts*—short ducts that pass over or through walls.

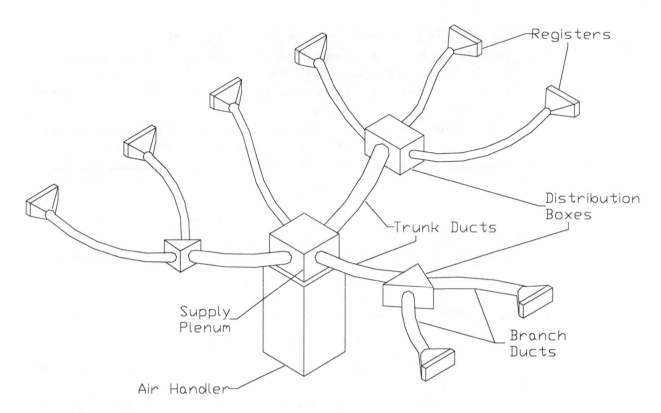

Figure 2.4: Spider System

Central and Multiple-Room Returns

A central return system consists of one or more large grilles located in common-use areas close to the air handler. Doors to individual rooms are undercut to allow return air to move toward the central returns. In multi-story homes, returns are located on each level, typically in a hallway.

Multiple-room return systems are designed and located to return the air supplied to each room. An improperly sized multiple return system will depressurize some rooms and pressurize others. Multiple-room returns are frequently installed using building cavities and stud spaces rather than "hard" ducts.

Central Returns

Advantages of central returns over multiple returns are:

- They require less ductwork as they usually consist of one large duct with a relatively short run.
- There is less friction loss for the same amount of air flow, thus minimizing blower requirements.
- Central returns are easy to install.
- They are often preferred with an open floorplan.
- Filter-grilles may be used to clean air before it enters the ducts.
- They have lower material and labor costs.

Some disadvantages of central returns are:

- They are generally noisier, unless special acoustical provisions are made.
- Doors to individual rooms must be undercut to permit proper air flow.
- Large ducts may require a special chase.
- Large grilles can be unattractive.
- Central returns can short-circuit air flow from nearby supply registers.

An example of a central return system is provided in Figure 2.5. Multi-level or two-story homes with central return systems would typically have a return air grille at each living level.

Multiple-Room Returns

Advantages of multiple-room (individual) returns include:

- Good sound attenuation from branch network, resulting in quieter operation.
- Improved air flow within individual rooms, even with doors closed.
- Better privacy, especially in bedrooms, since doors need not be undercut.
- Small branch ductwork, easily concealed within joist and stud spaces.
- Smaller, less conspicuous grilles.

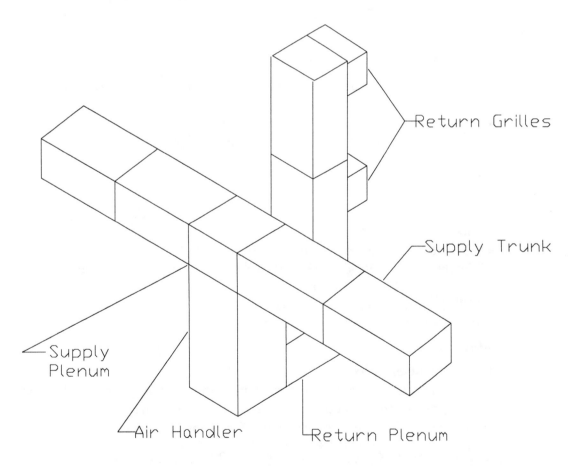

Figure 2.5: Central Return System

Disadvantages of individual returns include:

- They require an extensive duct system, usually trunk and branch similar to a supply system.
- Installation is more complex, usually requiring a separate layout.
- They are more costly to install than central returns.
- Improper sizing can still result in room pressure imbalances.
- They require a filter located near the equipment, which may be inconvenient for attic or crawlspace equipment and may require more frequent cleaning.

The choice of a return air system depends on performance criteria, construction limitations, and cost. Sometimes the best solution is a combination of two systems, employing a central return for large adjoining open areas and individual returns for smaller rooms such as bedrooms or dens.

Passive Returns

Passive return systems are added to central return systems to improve air movement between bedrooms and hallways when doors are closed. Returns may be formed using transfer grilles through partition walls, transfer grilles through closets or other connected spaces, crossover ducts placed in ceiling spaces, undercut doors, and door grilles or louvered doors. Figure 2.6 shows examples of passive returns.

Advantages of passive returns include:

- There is better return air movement between rooms and the air handler than with central returns.
- They are lower in cost than multiple-room returns.
- They can be used to enhance comfort in key areas of the home such as bedrooms when less costly central return systems are used.

Disadvantages of passive returns include:

- They may transmit noise and reduce privacy if improperly designed and located.
- They may not be aesthetically appealing.
- Architectural grilles, which help improve appearance, are more costly than standard painted metal grilles.
- They require some additional labor and material above central returns.

Panning

The technique of enclosing a floor joist or stud space for use as a duct is referred to as *panning*. Traditionally, sheet metal has been used for panning. However, drywall, hardboard, and other suitable materials also can be used. Panning is used primarily in the return air system. Its advantages include:

- It is allowed by residential mechanical code.
- Overhead space below the joists is not required.

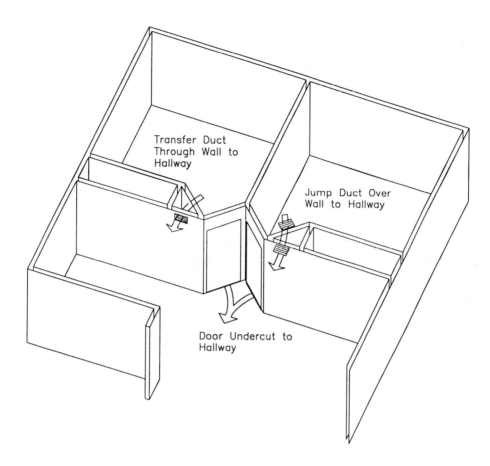

Transfer Duct
Through Wall to
Hallway

Jump Duct Over
Wall to Hallway

Door Undercut to
Hallway

Figure 2.6: Passive Returns

- Relatively large joist and stud spaces can provide ample air flow with minimal friction losses.
- Panning minimizes use of metal ducts and fittings, and work may be done by carpenters and other trades.
- It provides a low-cost multiple-room return system.

Disadvantages of panning include:

- Air leakage from panned and building cavity returns is comparatively high.
- Air leakage from cavities increases building infiltration rates, decreases overall energy efficiency, and may depressurize rooms or basement spaces, potentially resulting in health- or safety-related problems.
- Sources of cavity leakage air often are unknown and may carry pollutants from utility rooms or garage rooms.
- Sealing of building cavities is difficult, time-consuming, and largely ineffective.
- Sealing must be coordinated during framing, mechanical installation, and finishing.

Figure 2.7 shows combined use of a metal duct, a panned joist space, and a finished stud space for an air return. Any combination of these can be used.

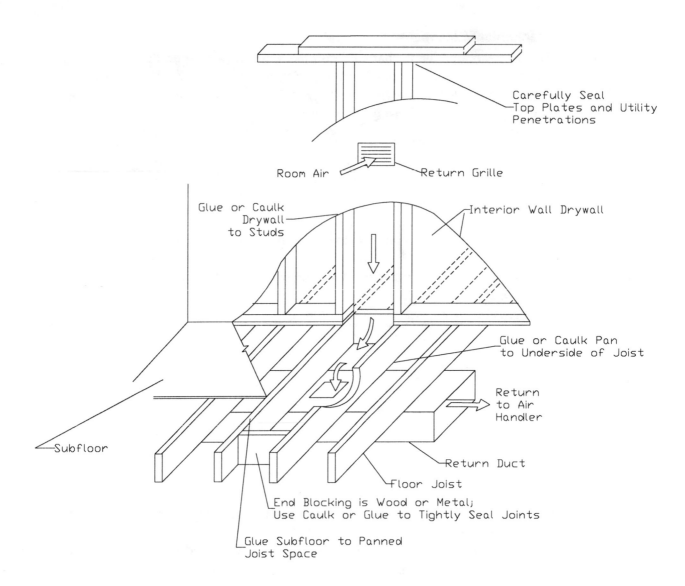

Figure 2.7: Panned Returns

Other examples of non-ducted air passageways include boxed-in chases, dropped ceilings (especially in hallways), and raised-platform equipment closets that use the enclosed space as a return air plenum.

Sound Attenuation

Sound attenuation is of particular concern in return air systems. Noise in a duct system usually can be traced to the return air ductwork, especially on systems with short central returns. These operating noises typically come from two sources:

- Mechanical equipment.
- Air turbulence.

Mechanical Equipment

Noises generated by mechanical equipment include air handler noise and mechanical vibrations. These sounds can be transmitted to the living space through the return ductwork. The more direct the sound path, the greater the noise level. Therefore, a multiple-return duct network will provide quieter operation. Less costly central return systems tend to transmit more noise.

Here are some tips for alleviating noise associated with central returns:

- Locate mechanical equipment and returns away from heavily used rooms or rooms where lower noise levels are desired.
- Set equipment on vibration pads, install flexible connections to ductwork, or use other similar techniques to isolate the mechanical system from the structure.
- Line the blower compartment and return duct with an acoustical duct installation.
- Provide at least one 90-degree elbow in return ductwork to establish an indirect path back to equipment.

Noise From Air Turbulence

Many of the techniques used to reduce mechanical noise also help reduce airflow noise. These techniques include acoustical linings and turns in the ductwork.

The most common sources of air turbulence noise are sharp changes in the direction of air flow in ductwork and excessive air velocity through grilles. Try using the following techniques to alleviate these problems:

- Size return ductwork and grilles adequately. Undersized ducts or grilles will result in increased noise levels because of higher air velocity and turbulence; ACCA *Manual D* prescribes allowable air velocities.
- Install fittings with a curved rather than an angular throat to provide smoother flow of air and to minimize turbulence and related noise.
- Install turning vanes or less costly (and less effective) splitters in the major elbows to promote uniform flow of air through fittings and across grilles. Air being drawn at high velocity through only part of the grille is a frequent source of noise. Although turning vanes are usually associated with commercial installations, they are also effective in residential work.

Chapter 3

Designing the Total HVAC System

This chapter describes the elements that make up a complete HVAC design and provides general guidelines for determining heating and cooling requirements, selecting HVAC equipment, and designing the distribution system. The information is intended as a guide to aid builders in understanding alternative systems so that you may deal more effectively with your subcontractors in obtaining a system that meets your requirements at the least cost. For precise system design, you'll still need the services of a competent HVAC contractor or a mechanical engineer.

DETERMINING HEATING AND COOLING REQUIREMENTS

As the builder, you will need to know the heating and cooling loads before either you or the HVAC contractor can select appropriate equipment and design the distribution system. These loads can be calculated by an HVAC contractor, a staff engineer, or an outside consultant. This must be done on a room-by-room basis to determine the amount of heat load and air flow that must be delivered to each room in order to maintain a well-balanced and comfortable temperature level throughout the house.

Heating Requirements

Loads calculated for the heating season are referred to as *heat loss* calculations, since heat is being lost from the building. Heat losses must be calculated for all portions of the building envelope, including walls, windows, doors, ceilings, floors, and foundations. Losses also are calculated for ductwork passing through unconditioned spaces and for air infiltration (building leakage). The total of these losses is expressed in Btu per hour (Btuh).

Loads are calculated for winter design indoor and outdoor temperatures. Winter indoor design temperature is typically 68°F. Winter outdoor design temperature is specified for a given location. Note that the design temperature is not the coldest outdoor temperature that may be expected. Actual temperatures may be below design temperatures 2½ percent of the time based on recorded averages. This data is presented by the Air Conditioning Contractors of America (ACCA) in *Manual J,* the residential building loads guide for cities and towns in each state. These temperatures are also available from the American Society of

Heating, Refrigerating, and Air-Conditioning Engineers (ASHRAE). Contact information for ACCA is provided in chapter 2. ASHRAE may be contacted at:

American Society of Heating, Refrigerating,
and Air-Conditioning Engineers (ASHRAE)
1731 Tullie Circle, NE
Atlanta, GA 30329
(404) 636-8400
http://www.ashrae.org

Cooling Requirements

Loads calculated for the cooling season are referred to as *heat gain* calculations. Heat gain is more complex than heat loss since, in addition to heat gains through the building envelope (similar to heat loss), there are gains from occupants, household appliances and lighting, and direct solar radiation through windows. The total of these gains can be sensed as heat by a thermometer or a thermostat, and are referred to as *sensible heat gain*.

In addition to the sensible heat gain, latent heat gain must also be calculated. Latent heat is the energy required to remove moisture from the building air. Calculations are based on an indoor relative humidity of 50 or 55 percent. Latent heat load can be significant—as much as one-quarter to one-third of the total cooling load in many homes.

Sensible gain plus latent gain equals the total heat gain for the structure. Similar to heat loss, total heat gain is expressed in Btuh. Loads are calculated for summer design indoor and outdoor temperatures. Summer indoor design temperature is typically 76°F. Summer outdoor design temperature is specified for a given location. As with heating, summer design temperatures are specified in ACCA's *Manual J* or ASHRAE's fundamentals. At each location, an additional piece of information is provided—the amount of moisture removed for each cubic foot of indoor air to reach 50 or 55 percent relative humidity. Heat gains are represented as tons of cooling, one ton equaling 12,000 Btuh.

Example: A house has a calculated total heat gain of 29,430 Btuh:

$$\frac{29{,}430 \text{ Btuh}}{12{,}000 \text{ }^{Btuh}\!/_{ton}} = 2.45 \text{ tons} = 2\frac{1}{2} \text{ tons (nominal)}$$

Estimating Heating/Cooling Loads

Several procedures are available for the calculation of heat losses and gains. See *Manual J*, Load Calculation for Residential Winter and Summer Air Conditioning developed by ACCA. Whichever calculation you use, you should require that room-by-room heat-loss and heat-gain calculations be submitted for each house type; don't settle for a short-cut calculation or educated-guess method. If large unshaded glazing areas are incorporated into the architectural design and a standard house type is to be repeated on many different orientations, a separate calculation for each major orientation will be needed.

After some experience in reviewing heating/cooling loads, you may wish to develop a reference table summarizing Btuh-per-square-foot that applies to similar house types for the local climate. While approximate loads of this type do not provide a basis for designing the HVAC system, they do help you to develop a feeling for the heating and cooling loads and enable you to evaluate the feasibility of an HVAC proposal. By comparing approximate numbers of this type with calculations in a proposal, you can easily identify a major load calculation error. You may also use these approximate numbers to roughly assess energy and equipment requirements on proposed house designs. Be careful to compare homes with similar window areas, insulation levels, and infiltration rates.

SELECTING HVAC EQUIPMENT

This section describes different types and configurations of equipment, alternate fuels, and comparative installation and operating costs. General equipment characteristics are then summarized to provide a basis for selecting the equipment.

Correctly sized equipment and ductwork are important for good system performance. If cooling equipment is undersized by more than 15 to 30 percent, a temporary temperature rise (temperature swing) of 4.5 to 6°F can be expected above the indoor design temperature when the outdoor design conditions are exceeded. This compares to a temperature swing of 3°F when the unit is exactly sized. If oversized by more than 10 percent for both heating and cooling, the equipment will cost more initially and will short cycle (i.e., have very brief run times), resulting in lower comfort levels, higher operating costs, and wasted energy. In addition, oversized cooling equipment may not provide adequate moisture removal because of reduced running time.

Designers have a tendency to oversize HVAC systems. They round their calculations to the high side, add a percentage for safety, and then select the next-larger-size equipment. Oversizing is not necessary because even a properly sized system is over-capacity except when operating at design conditions.

Types of Heating and Cooling Equipment

Several types and combinations of heating and cooling equipment are available. Your selection will depend on available fuels, building design requirements, and other factors. The most common types of HVAC equipment are discussed in the following sections.

Gas or Oil Furnaces

Furnaces are available in four basic configurations to allow for a variety of applications and airflow directions: upflow, lowboy upflow, counter- or downflow, and horizontal flow. Furnaces are rated in units of AFUE (Annual Fuel Utilization Efficiency). The higher the AFUE, the more efficient the equipment. Oil-fired units are customarily installed in basements, crawlspaces, or indoor utility spaces. Attic installations are not performed, since two-stage oil pumps are required to lift the heating oil and because of the potential for damage or odors from oil leaks.

Upflow furnaces are used where supply ductwork will be installed above the unit. The furnace may be located in a basement with the distribution system just below the first floor joists, or in a first-floor utility room with the distribution overhead in an attic or dropped ceiling. Upflow furnaces are usually available in more sizes and at less cost than other types of units. A reduced-height lowboy furnace is available and is similar to highboy (or standard) upflow units but can be used where ceiling heights are restricted. Typically installed in basements of older homes as replacement units, they are generally more costly than highboy units.

Counterflow furnaces are frequently installed in single-story homes built on a slab or crawlspace where the air distribution system is below the floor. Counter-flow furnaces are usually more costly than upflow units. When installed on wood-floor systems, most models require a non-combustible floor base which is purchased as a furnace accessory.

Horizontal furnaces are designed primarily for installation in an attic or crawlspace where both the unit and the distribution system will be located. Horizontal furnaces are more costly than upflow or downflow models.

Wood Pellet and Coal Stoves
The old, smoky, wood- and coal-burning stoves have been redesigned into efficient heating sources which are finding increasing use in vacation and weekend homes. Most units have variable air flow controls, optional thermostats to control burn and heat supply, and battery backup controls to allow firing during power outages.

Electric Furnaces
An electric furnace is simply an air-handler unit with electric resistance heating elements inside. Units are sized by Btuh or kilowatts (kW), 1 kW equaling 3,412 Btuh. Resistance elements are typically available in 5-kW increments. Electric furnaces are compact and can be used in upflow, downflow, and horizontal installations with little or no modification. They have a lower first cost than fossil furnaces, but operating costs are generally higher because electricity can be more expensive than other fuels in many regions. Resistance elements convert 100 percent of supplied electric load into heat.

Split-System Cooling Equipment
When cooling is incorporated with a standard furnace, a split system is used. A split system consists of an outdoor condensing unit containing the condensing coil and compressor, the indoor evaporator coil (or A-coil), the interconnecting refrigerant tubing, and a blower which is part of the furnace. Cooling capacities

of standard residential equipment range from 1½ to 5 tons. Performance efficiency of cooling equipment is factory-determined under a specified set of test conditions and rated as SEER (Seasonal Energy Efficiency Ratio). The higher the SEER, the more efficient the equipment.

Heat Pumps

Two types of heat pumps are used in residential work: air-source (air-to-air) and ground-source (water-to-air). Air-to-air heat pumps are more common and resemble split system cooling equipment, except that coolant direction is reversed for heating. An electric furnace is generally used to provide air handling and supplemental resistance heat, but fossil-fueled equipment can also be used. Ground-source heat pumps use groundwater or earth temperature for a heat sink. Since ground temperatures do not change much over the year, ground-source heat pumps have higher seasonal efficiencies than air-source heat pumps.

The advantage of a heat pump is that both space heating and cooling can be delivered for only a few hundred dollars more than air conditioning alone. Also, a heat pump is two to three times more efficient than electric resistance heat. Efficiency of an air-to-air heat pump drops with dropping outdoor temperature. In most systems, when the heat pump can no longer meet demand, heating is supplemented by electric resistance elements. Air-to-air heat pumps are best suited to areas with relatively moderate winter and summer weather and least suited to areas with extremes of either cold or hot weather, since both heating and cooling are provided by the equipment. Oversizing of either cycle should not exceed 25 percent. Heat pump heating cycles are rated in HSPF (Heating Seasonal Performance Factor). The higher the HSPF, the more efficient the equipment.

Self-Contained Units

Cooling equipment is also available in a self-contained unit in which the condenser, blower, and coils are included in a single package. A heating option, usually gas or electric, is generally available on this type of equipment. Heat pumps are also available as self-contained units. This equipment can be either roof-mounted or located at ground level outside the house. Neither floor space nor interconnecting lines are required, and the cost is normally less than it is for a split system. Self-contained units are generally used in areas where cooling predominates. Typically, rooftop units are installed in conjunction with an insulated attic duct system. Units may also be set on a ground-level concrete pad to supply a crawlspace or basement duct system. One disadvantage of self-contained units is that operating costs may be somewhat higher because the air handling equipment and some ductwork are located outside of the conditioned space.

Installation and Operating Costs

Both initial installation cost and operating cost are important considerations when you select equipment. Operating cost depends on the local availability and cost of fuel, and on the seasonal efficiency of the equipment. The primary heating fuels are natural gas and electricity. Less common are #2 heating oil, liquid propane (LPG), coal, and wood. For cooling, electricity is predominantly used.

DESIGNING THE DISTRIBUTION SYSTEM

Once you have determined heating and cooling loads and have selected the type of equipment to be used, you can design the distribution system. Your primary design concern will be proper sizing of the ductwork. For optimum performance, the ductwork must be sized to handle the airflow delivered to each room by the HVAC equipment. Two sources you may use for sizing ductwork are ACCA's *Manual D: Duct Design for Residential Heating and Cooling*, and SMACNA's *HVAC Duct System Design*. Contact information for ACCA and SMACNA are provided in chapters 1 and 2.

Determining Equipment Airflow

The total airflow that a piece of equipment delivers to the distribution system is what essentially determines the size of the ductwork. Airflow is specified in cubic feet per minute (cfm). The ducts must be able to pass the required air flow and direct it to the conditioned spaces. Air handlers usually have several speed settings available; they also have curves which specify the delivered air flow for a given amount of duct and equipment resistance. For precise airflow determination, consult the manufacturer's literature for specific equipment. Such literature usually includes allowances for various system friction losses such as coils and filters and for operating at various static pressures.

When heating and cooling are combined in the same system the ductwork is sized for the operating mode which requires the highest air flow; this is usually cooling. Manufacturer application data for air conditioners can range from as low as 320 cfm to as high as 700 cfm. Equipment rating data lies between 350 and 450 cfm. As a rule-of-thumb, air conditioners are typically specified for duct systems which use 400 cfm of air per ton of cooling. Gas and oil furnaces are typically designed to raise air temperature between 35°F and 60°F, which results in 180 to 312 cfm per ton of heating. Fan speed and equipment capacity must be negotiated between these somewhat different requirements.

Supply and Return Locations

The location of supply and return outlets can have a significant effect on the performance of an HVAC system. An understanding of air flow characteristics and common sense are necessary for effective placement of registers and grilles. Outlets and returns should be located to provide an acceptable level of air mixing in the most critical operating mode, whether heating or cooling.

Supply and Return Air Patterns

A supply outlet is responsible for most of the air movement within a room. The distance it can deliver its airstream is referred to as *throw*. In addition to throw, supply air has the ability to induce motion in otherwise stagnant room air by entrainment into its main airstream. Similar to the suction end of a vacuum cleaner hose, the return air intake can only collect air in the immediate vicinity of the grille. This localized influence is in direct contrast to the airstream issuing from a supply outlet.

Keep these factors in mind when determining the location of supply outlets and returns:

- Air patterns within a room are primarily determined by placement and sizing of supply air registers.
- Low return air intakes draw cooler air from the floor to improve mixing of room air during heating system operation.
- High return air intakes draw warmer air from the ceiling to improve cooling.
- Air patterns are also affected by drafts and thermal air currents.

Air Stratification

Air stratification is another factor affecting register and grille placement. Stratification is the tendency of different layers of temperature to develop from floor to ceiling. In older homes having low levels of thermal protection, the temperature difference between the floor and ceiling may be 5 to 10°F or more. The placement of registers and grilles in such homes was considered critical in reducing these differences. However, in most houses built with today's higher levels of thermal protection, temperature stratification is in a more acceptable range of 2 to 5°F, and register and grille placement is much less critical. Also, less stratification occurs with modern equipment because the delivered air temperature is lower.

Supply Locations for Year-Round Operation

Supply outlets are traditionally located near the source of highest heat loss (heating) or highest heat gain (cooling). Supply outlets are generally placed to serve the more critical need since it is impractical to install two separate supply systems.

In cold climates where heating requirements prevail, perimeter floor outlets are generally preferred. In hot climates where cooling prevails, ceiling diffusers or high wall outlets that blanket the entire area are used to provide good cooling distribution.

In more moderate climates, particularly in well-insulated homes, outlet location is less critical. Supply registers located along interior walls are satisfactory and result in a significant reduction in ductwork. An adjustable straight-blade register sized to throw the conditioned air across the room provides adequate air mixing for cooling and heating. Ducts in outside walls should be avoided because they are difficult to insulate adequately.

Return Locations for Year-Round Operation

While return air grilles have significantly less effect on room air motion than supply air registers, they can still be located to enhance total system performance. In heating, since warm air rises, a low return helps the supply outlet perform by drawing off the cooler air near the floor. In cooling, since cool air falls, a high return draws off warmer air near the ceiling. For year-round operation, returns should be placed to serve the more critical need, as follows:

Operation	Return Location
Heating only	Low
Cooling only	High
Heating predominant, some cooling	Low
Cooling predominant, some heating	High

On multilevel and two-story homes with both heating and cooling, good judgment again applies. Since upper levels will tend to heat more readily by natural convection, cooling is more critical and the upper-level returns should be placed high. Conversely, lower levels will tend to cool by convection, and the returns should be placed low to improve heating.

Placing returns in some rooms or areas may be ill-advised or restricted by building code for reasons of odor, safety, or comfort. These areas include:

- Bathrooms.
- Kitchens.
- Garages.
- Mechanical equipment rooms (if fossil fuel).
- Areas where noise or drafts may be objectionable.
- Rooms with natural draft fireplaces.
- Basements.

Ductwork Sizing

Proper duct sizing allows for optimal performance within the limitations of the system for which it was designed. Undersizing can result in higher noise levels, higher operating costs, reduced comfort, and resulting consumer complaints. Oversizing of supply ductwork can result in system imbalance and higher installation costs.

Total air-carrying capacity of a supply duct depends on the cross-sectional area, total length, friction losses in ducts and at fittings, type of duct material, and blower output. When duct systems are undersized the system will operate less efficiently, run at higher noise levels, and provide less effective conditioning. A larger size promotes efficiency by reducing internal air friction, air velocity, and blower horsepower requirements. This tends to improve operating characteristics and reduces the noise generated by air turbulence.

Chapter 4

Energy-Efficient Practices

Energy-efficient construction is a blend of basic energy principles and good construction practices. As a builder, your role is to mediate between efficiency and constructability, in order to provide a cost-effective, efficient, and comfortable home. With a basic understanding of energy-efficient principles and an understanding of the interdependence of each building component, you may properly participate in the design and installation of cost-effective and energy-efficient HVAC systems and building components. Energy-efficient principles and construction practices will be discussed in this section. The benefits of these practices include: lower heating and cooling loads, smaller equipment, smaller and simpler ductwork, lower utility bills, and increased home comfort.

It is important to keep four key ideas in mind:

- Properly size and install equipment and ductwork. Efficient and accurate design will provide cost savings to you and improved comfort and utility savings to the homeowner.
- Reduce duct leakage and improve duct insulation. Significant duct system losses are due to leakage of conditioned air from the supply ducts and into the return ducts from unconditioned spaces and to conduction of heat between the duct system and its environment. Ducts should be tightly constructed and properly insulated, and you should inspect them to verify quality.
- Ducts should be designed into conditioned spaces to the greatest extent possible. The cost of moving ducts out of attics, crawlspaces, and unconditioned basements can be offset by smaller equipment, smaller or less ductwork, reduction of duct insulation, and reduced emphasis on duct sealing. The significant benefit is a more energy-efficient home with little or no impact on construction costs.
- Improve overall HVAC system efficiency. This can be done with trade-offs between more efficient duct systems and more efficient equipment.

PROPERLY SIZE AND INSTALL DUCTS AND EQUIPMENT

Proper sizing and installation of ductwork and equipment provides better, controlled comfort and allows smaller equipment to be specified. To accomplish this, you should understand and know the peak summer and winter building loads, envelope infiltration, and duct leakage for the proposed building. These loads

32

and air leakage quantities are used to properly size heating and cooling equipment for new homes. Load calculations are easily performed by large production builders or luxury custom builders. Production homes are repetitive and the added cost of calculating loads and equipment sizes can be spread out over a large number of homes. Luxury homes focus on occupant comfort and buyers will likely appreciate the additional cost of properly sizing equipment and ductwork.

Building Load Calculations

Energy may be lost from ceilings, walls, foundations, windows, ducts and equipment, and infiltration. To accurately size equipment for peak space conditioning loads, all that you need are basic dimensional information and insulation values. Dimensional information includes floor and ceiling areas, net wall area (without windows and doors), window and door areas, and heated perimeter. Insulation values are R- or U-values and any adjustment for framing. Design indoor and outdoor temperatures are also required.

Specify Design Temperatures

Be sure to specify indoor and outdoor design temperatures to the HVAC contractor in order to prevent equipment oversizing and to get bids based on the same parameters. Typical winter and summer thermostat settings are 68°F and 76°F, respectively. Outdoor design temperatures are published in many documents.

Ceiling, Wall, and Foundation Losses

Losses through ceiling, wall, and foundation systems typically comprise a third or more of the building load. Ceilings are increasingly insulated to R-30 or better, which substantially reduces attic impact on heating and cooling losses. Conventional 2x4 walls are limited to R-11 to R-15 for batts. Alternatives may give insulating values of as much as R-23. Basements in new construction are increasingly being insulated with R-10 wall blankets.

Window Losses

Energy-efficient, double pane windows have U-values that range from lows below U-0.35 (R-2.9) to highs above U-0.80 (R-1.25). Typical walls and ceilings are R-13 and R-30, respectively. Therefore, conduction losses can be more than 10 times higher for windows than for walls and more than 14 times higher than for ceilings on a square-foot basis. It is important to know the R-value of windows selected and to know the exact footage of window area for calculations.

In summer, the orientation of windows has a dramatic effect on cooling requirements. Just a 90-degree shift in building orientation could save half a ton or more on air conditioning equipment. You should know the load on your designs for each orientation. This will reduce equipment sizes, promote proper duct sizing, improve homeowner comfort, and reduce homeowner utility bills.

Duct Losses

Duct losses can be a large component of heating and cooling energy use and must be considered in order to better define the conditioning requirements of a new home. Ducts may be placed in conditioned spaces to eliminate duct efficiency concerns and to improve energy efficiency and overall system capacity by as much as 30 percent.

Equipment Losses

Equipment should be properly matched to required loads. Grossly oversized equipment runs in short bursts or cycles and rarely reaches steady operating conditions. It also has a greater chance of premature failure. This part-load or short-cycling condition reduces equipment efficiency and durability. Also, equipment costs increase directly with increasing capacity.

Manufacturers and HVAC trade organizations specify equipment size compatibility. Air conditioning coils require system flows of 400 cfm per ton of cooling. Mixed furnace and air conditioner systems must match air flow from the furnace to the air conditioner requirements. ACCA recommends a maximum of 40 percent oversizing of gas-fired furnaces when coupled with air conditioning coils. Also, air conditioners and heat pumps should not be oversized by more than 25 percent of required sensible cooling load.

Infiltration Losses

In new construction, building infiltration is typically one-fourth of the building heating and cooling load. In the winter months, wind and temperature effects cause increased leakage of warm, conditioned air to leave through window and door seals, band joists, duct and fireplace penetrations, and electric and water penetrations. In the summer months, infiltration brings moisture. Moisture is removed only through longer equipment run times. To combat infiltration loads, larger, more expensive equipment is often used to provide the same comfort to the homeowner. But a more efficient approach is to tighten building envelopes by using caulks, adhesives, and air barriers.

Also, you need to know the building's infiltration rate. Simple blower door tests of a few homes give a good indication of the leakiness or tightness of a builder's housing stock. Accurate infiltration rates should be used to more precisely size winter and summer design loads and equipment. Accurately sized equipment saves first cost, improves energy bills, and improves home comfort. The following table gives an indication of home infiltration rates as determined by a standard "blower door" test.

Measured Leakage	Very Tight	Tight	Average	Leaky
Air Changes at at 50 Pascals, ACH50	3 or less	3 to 6	6 to 10	10 or more

A home can be made "very tight" through use of alternative construction materials and extreme infiltration measures. A "very tight" home will require external ventilation, usually heat recovery ventilators or intermittent ventilation fans. A home may be made "reasonably tight" by targeting key areas such as band joists, wall plates, corners, and window and door openings with construction adhesives, caulks, and air barriers. A "tight" home may require additional ventilation. A home constructed to an "average" level of leakage will use few, if any, infiltration reduction measures. Average homes do not require additional ventilation. Leaky homes are usually old homes or poorly constructed new homes, and they have large infiltration losses. Leakage paths include panned returns, ducts in unconditioned spaces, bandjoist leakage, and top and bottom plate leakage.

Duct Layouts and Sizing

You and the HVAC contractor should plan out duct systems on a set of drawings; the drawings should then be submitted to you with accurately calculated duct and equipment sizes. Three goals should be established:

- Keep the duct system inside. Don't locate ducts in unconditioned spaces such as attics, crawlspaces or garages. Do place ducts in interior areas, such as floors and interior walls, that are within the conditioned space of a home.
- Keep it tight. Don't install leaky duct systems in unconditioned spaces such as attics and crawlspaces. Do seal ducts located in unconditioned spaces with appropriate tapes and mastics.
- Keep it simple. Don't install ducts with numerous joints and turns because of unexpected framing conditions resulting from poor planning and coordination. Do install simple duct systems with straight runs and few joints.

REDUCE DUCT LEAKAGE AND IMPROVE DUCT INSULATION

Duct losses are due primarily to duct leakage and conduction losses. Duct leakage can be controlled through improved construction and verification of proper installation. Insulation can be improved by properly sizing insulation for each system.

Duct Leakage

Forced-air distribution systems are constructed from a variety of duct materials, fittings, and components which create many joints and seams. Small openings in these connections leak air during air handler operation because of pressure differences across the duct wall. Leakage of conditioned air from ductwork results in energy losses and reduced occupant comfort. Duct leakage can be reduced through better construction and sealing methods, innovative duct products, and more rigorous inspection of finished work.

Air leakage from a duct system is probably the single largest cause of duct inefficiency. In heating, air within ducts is typically between 30 and 60 degrees higher than house temperature and typically more than 70 degrees higher than outdoor air. In cooling, conditioned air is more than 15 degrees cooler than house temperature, 35 degrees cooler than outdoor air. In addition, summer ambient air has moisture or humidity that must be removed from the air to maintain home comfort. The loss of conditioned air from the duct system to unconditioned spaces results directly in lower distribution efficiency. Supply leakage loses conditioned air which has temperatures much higher or lower than outdoor temperature. Return leakage pulls outdoor air into the duct system, adversely affecting air temperature and humidity. Low leakage is especially important for ducts placed outside of a home's conditioned space.

Tighter Construction

Duct construction can be improved by adhering to industry standards and adjusting the order of construction. Consult industry standards to determine the best approaches to constructing duct systems. Substandard construction practices are used by some HVAC contractors because of inexperience, poor craftsmanship, or to reduce costs. Whatever the reason, knowledge of standard practices allows you to take a proactive role in HVAC installation.

In some cases, the order of construction affects installation quality. Ducts should always be placed before plumbing and electrical systems are installed. Ducts require more room and cannot easily be moved. Ducts should also be designed with consideration of foundation and framing details. Chases and joist directions should be arranged to reduce changes in duct direction and numbers of connections. Duct systems should be visualized in three space dimensions to foresee any conflicts between framing and HVAC. Further, if sealing is desired, ducts should be constructed with sealing in mind. Trunks cannot be sealed properly when placed side-by-side within a chase or joist cavity or placed tightly against sub-floor materials. It may be possible to construct the trunks in large sections, sealing joints while accessible, and slipping them into place.

Duct Sealing

Duct leakage can be reduced by sealing ducts during and after their placement. Aluminum-backed tape, mastic, or heavy butyl-backed tape are used in sealing ducts. It is advisable, however, to establish how each connection will be sealed and to gauge the effectiveness of the sealing strategy.

Leakage reduction is most effectively accomplished if the duct system is designed with sealing in mind. Trunks should not be placed back-to-back in the same chase or tightly against ceilings or floors. If these designs cannot be avoided, ducts should be substantially sealed prior to installation. Prioritize duct connections to be sealed. Leaks nearest to the air handler have the greatest pressure difference or driving force and will leak more than a similar leak closer to the registers. Techniques for sealing each type of duct connection should be specified and construction should be confirmed through visual inspection and testing.

Ducts are difficult to seal once placed overhead in basements, under floors in crawlspaces, or placed in attics. Sealing is accomplished more effectively and inexpensively during construction. Building cavities should not be used, but if doing this is unavoidable, they should be sealed during construction. Construction adhesives or gasket materials can be applied to studs and floor joists before placing sheet rock or plywood. This requires coordination and training of subcontractors. In production homes it becomes a detail to track to completion.

Leakage is measured using a fan attached to the duct system (see Testing, below). The following table suggests a guideline for gauging the level of tightness of a duct system:

Table at 25 Pascals

Leakage Level	Unconditioned Leakage (cfm at 50 Pa)	Total Leakage (cfm at 50 Pa)
Very Tight	<30	<130
Tight	30 to 100	130 to 250
Average or Typical	100 to 150	250 to 500
Loose	>150	>500

Table at 50 Pascals

Leakage Level	Unconditioned Leakage (cfm at 50 Pa)	Total Leakage (cfm at 50 Pa)
Very Tight	<50	<200
Tight	50 to 150	200 to 400
Average or Typical	150 to 250	400 to 800
Loose	>250	>800

Testing

Diagnostic tests of newly constructed duct systems may be used to identify leakage sites and to evaluate and compare leakage rates between systems. Pressure tests may also be used to confirm tight construction and effective sealing measures. Tests are performed by pressurizing duct systems at a return register or the air handler cabinet. Duct pressure may be measured at any location, but you should make an effort to measure an average or representative pressure across all of the duct leaks. Taps should be placed in a consistent manner between houses tested.

Duct leakage can be separated into two types: unconditioned and conditioned leakage. Unconditioned leakage is lost from the building envelope and is an unrecoverable energy loss. Leakage to conditioned spaces is regained by the home but may result in reduced comfort. Energy contractors should be able to evaluate either of these leakage types.

Inspection

Typical duct system problems include disconnected ducts, pinched or crushed ductwork, missing or badly torn duct insulation, or poor duct layouts. As the builder, you should specify the duct materials, equipment, and installation practices to be used in a new home, and you should inspect the completed HVAC system against the contracted specifications to determine compliance. Sample inspection checklists are provided in chapter 5.

Duct Insulation

Duct insulation should be inspected for proper installation. For duct running through unconditioned spaces, conductive loss is reduced by using code-required or greater insulation levels.

Minimum Thickness Levels

Minimum insulation values are specified in model building and energy codes. The 1995 CABO 1 and 2 Family Dwelling Code, for example, specifies a minimum of 2 inches of 0.75 pound density fiberglass or mineral wool insulation or 1 inch of 1.5 pound density duct liner for metal duct systems placed in unconditioned spaces. The code further specifies a minimum R-value of 4.2 for non-metallic ducts.

The 1995 Model Energy Code, for example, specifies insulation R-values based on design temperature differences. *Temperature difference* (TD) is defined as the absolute value of the difference between duct air temperature and duct ambient air temperature. For a TD of less than 15°F, no insulation is required. For a TD of between 15 and 40°F, insulation with an R-value of 3.3 is required. And for a TD of more than 40°F insulation with an R-value of 5.0 is required. Duct air temperature will vary with distance from the air handler and will fall somewhere between room temperature and plenum temperature. Plenum temperatures for gas and oil furnaces are typically 35 to 60°F above room temperature and are 20 to 30°F above and below room temperature for heat pumps and air conditioners, respectively.

Consult your local building code to determine minimum insulation levels. Insulation for ducts is most effective when properly installed. Gaps or tears in insulation aggravate heat transfer and can allow condensation to occur. Gaps and tears in duct insulation should be covered with aluminum-faced tape or additional insulation.

Maximum Levels and Diminishing Returns

Energy savings of 1 to 2 percent are gained for a typical home by increasing attic insulation from R-4 to R-6 and another 0.5 percent savings is gained from increasing insulation from R-6 to R-8. Costs for upgrading insulation thickness increase much faster than energy savings. Just increasing from R-4 to R-6 insulation increases material prices by 22 percent to 45 percent, depending on the type of insulation or ductwork. Increased insulation costs should be objectively compared to potential energy savings on an individual duct system by duct system basis.

Also, design tools such as ACCA's *Manual J* and *Manual D* do not account for duct lengths. You may want to add additional insulation to long duct runs to keep delivery temperatures comfortable.

MOVE DUCTS INTO CONDITIONED SPACES

Conditioned space may be defined as an area or volume of the home that is deliberately heated and cooled or a space that is adequately insulated against outdoor heat loss and uses recovered heating or cooling. Ducts and equipment placed within conditioned spaces are more efficient than those placed in unconditioned spaces. Conduction and radiative losses, leakage losses, and equipment cabinet losses are reduced or regained into the building space. Placement of ducts within conditioned spaces has the single largest impact on the energy efficiency of modern homes. The energy savings are similar to those provided by the use of expensive, high-efficiency equipment.

The following table shows the impact of duct location on distribution efficiency.

Location of Ducts	Heating	Range of Duct Efficiencies	Cooling	Range of Duct Efficiences
Conditioned Space	BEST	85 to 100%	BEST	90 to 100%
Attic	AVERAGE	60 to 80%	WORST	60 to 80%
Crawlspace	AVERAGE	65 to 85%	AVERAGE	70 to 90%

The following sections present techniques that may be used to bring ductwork and equipment into the conditioned space of a home.

Basement Homes

Basement homes have room to run the main trunk and plenum across the ceiling of the basement, perpendicular to the direction of the floor joists. Ducts in attics may easily be moved into conditioned space by running vertical risers off of the main trunk through interior walls and partitions. Typically 3x10 or 3x12 ducts will be run straight off of the basement trunk directly to the second floor. From there they will branch to the front and back of the building to supply the rooms in those areas. Six or more risers might be needed to feed six rooms on the second floor.

A second approach is to use floor truss or I-joist systems. Floor trusses allow a central trunk to pass down through the center of the truss or for take-off runs to pass easily through openings. I-joist systems allow for fair-sized cut-outs through the spans. The size and location of these cut-outs must be coordinated with the I-joist manufacturer. Be careful when designing duct systems that run between floors in homes with two-story foyers and family rooms. It may be difficult to supply and return air from these areas.

Crawlspace Homes

Equipment in crawlspace homes is typically placed in the garage, crawlspace, or equipment mechanical room and ducts are placed in the crawlspace. Sometimes both equipment and ducts are placed in the attic. Design trade-offs must be made to ensure space within the conditioned space of the home for ducts and equipment. A good place to start is with alternatives to dimensional lumber. Using I-joists or trusses on only the second floor will allow ducts to be run perpendicular to the floor joists. Ducts and equipment can be substantially downsized because leakage and conduction losses will be regained by the building space of the home. Equipment could be placed in a closet on the first or second floor. A single-story crawlspace home is essentially treated as a slab home. The disadvantages of moving equipment and ducts into the conditioned space are lost floor space and equipment noise.

Slab Homes

In slab homes, ductwork and equipment are typically placed in the attic. Again design trade-offs must be made to ensure space for ducts and equipment within the conditioned space of the home. For a single-story slab home, ducts may be placed within the slab or within interior soffits and dropped ceilings in hallways. For ducts in the slab, plastic duct products provide an alternative to coated sheet metal systems and have little or no leakage. Slab edge insulation should be used to prevent heat loss and to create an improved radiant-slab effect. Downsized ducts and equipment may be used to offset the cost increase caused by moving ducts inside.

IMPROVE SYSTEM EFFICIENCY

System efficiency is the combined efficiency of HVAC equipment and ductwork and is essentially the product of equipment and distribution efficiency. System efficiency may be increased by increasing equipment efficiency or by increasing distribution efficiency or by improving both.

Case Comparison	Equipment Efficiency		Duct Efficiency		System Efficiency	Cost Considerations
Case 1: High efficiency gas heating equipment and all ducts in unconditioned space	92%	x	70%	=	64%	Cost of high efficiency equipment versus standard efficiency equipment
Case 2: Standard efficiency gas heating equipment and all ducts in conditioned spaces	78%	x	98%	=	76%	Cost of moving ducts to interior less the cost of no duct insulation, reduced equipment size, and reduced duct sizes

Similar estimates may be made for air conditioners or heat pumps. Upgrading an air conditioner from a Seasonal Energy Efficiency Ratio (SEER) of 10.00 to a SEER of 12.00 increases equipment efficiency by 20 percent and increases equipment cost by as much as 65 percent. Upgrading an 80 percent AFUE (Annual

Fuel Utilization Efficiency) furnace to a 92 percent AFUE furnace improves equipment efficiency by 15 percent and increases equipment cost by as much as 70 percent. A better price trade-off could be made by increasing duct efficiency. A duct improvement needs to be applied only once, and improves both winter and summer operation. For combined furnace/AC systems, both pieces of equipment would need to be improved to get the same benefit.

Chapter 5

HVAC System Quality

Duct systems in residential construction are the product of proposals, contracts, planning, design, material specifications, and installation. The factors that determine the quality of a forced-air distribution system are the same as for many other consumer goods, such as cars and appliances. As a builder, you need to understand both technical issues, such as how well ducts are constructed, and non-technical issues, such as the selection of a good HVAC subcontractor, to effectively manage HVAC system quality.

Some recommendations in this chapter involve very little effort to implement, such as including inspections of ductwork at various stages of the construction process. The purpose of this document is not to teach the details of quality management. Rather, some practical tips for HVAC contractor selection, proposal evaluation, the HVAC contract, planning, coordination, and inspection are given.

DEFINING DUCT QUALITY

As a quality builder, you should begin with a clear definition of what quality means for the residential ductwork and HVAC systems that will be included in a given home. Four likely candidates for quality goals are:

- Minimizing defects or contract deviations.
- Maximizing comfort.
- Maximizing energy efficiency (thereby minimizing operating cost).
- Minimizing first cost.

In some cases, additional quality goals, such as air quality and noise, may be important. These goals are easy to list but much harder to quantify, particularly when trying to balance competing interests such as maximum energy efficiency and minimum first cost. This situation is where an effective quality process works best and, when implemented, will naturally result in construction that comes within practical limits of the overall quality objectives.

The first item listed above (minimizing defects) relates to the construction phase, and may be managed with the use of effective inspection and defect-correction procedures at the jobsite. The last three items in the list are design issues, which establish the level of quality well before construction begins.

SELECTION OF A GOOD HVAC CONTRACTOR

Since a builder is a manager of both home construction and design, the best way to get quality HVAC systems and ductwork is to select a good HVAC contractor and make sure your quality goals are well understood. A good HVAC contractor will be able to propose systems (sometimes with various options) that will allow the builder to select the best trade-offs between low first cost, energy efficiency, and competing quality goals.

The approach to selection of a good HVAC contractor is no different than that used for the purchase of many other types of services. The selection process should include:

- Checking references—measuring the candidate's ability to "get along" and perform.
- Cross-evaluating bids—measuring cost competitiveness.
- Evaluating design capabilities—measuring the candidate's ability to accurately size and meet performance criteria for HVAC systems.

Many other contractual issues related to liability and insurance are also important to consider, but the items above deal primarily with conditions that will affect HVAC system and ductwork quality. Additional information on the selection of trade contractors is available from other NAHB and NAHB Research Center resources.

The Bid

The selection of a good HVAC contractor is largely a function of how well the builder creates and evaluates requests for bids. You should have a clear picture of your quality goals. These goals should be a companion to first cost in evaluating bids.

The Scope of Work Statement

When developing a bid, it is important to provide the information that the HVAC contractor will need to perform an accurate preliminary design and cost estimate. The scope of work must be clearly defined, including a detailed house plan with specifications related to building energy performance (i.e., insulation levels, window types, house orientation, landscaping, etc.). Specific interests in alternatives, such as high-efficiency equipment or programmable controls, should also be listed. Material and installation specifications should be either referenced in the bid or requested in responses to the bid. Any special requirements such as duct sealing should also be identified. To the extent possible, the scope of work should include performance objectives that relate to the quality goals for the HVAC system and ductwork. The bid should basically create a "level playing field" from which to evaluate HVAC contractors. If it does not, you may be duped and your customer is the one who will probably suffer the most.

Developing a scope of work statement for HVAC work is not an easy task for a builder who knows little about HVAC systems and energy performance. Therefore, it may be worth your investment to develop a general scope of work with a knowledgeable consultant or HVAC contractor. This general scope of work may be modified and applied to many future projects.

Evaluating Responses to the Bid

In general, responses to a bid should be evaluated for completeness, competency, and cost. You should be looking for the best HVAC contractor with the best price. In some cases, HVAC contractors may provide additional information or recommendations that improve the value of their work and your homes. These types of responses should be encouraged and carefully considered. Value-engineering of the HVAC system may be included in the bid as part of the planned scope of work in the future contract agreement. In this case, the bid is evaluated with an emphasis on technical capability as well as cost. This approach can pay off in lower first costs, more efficient HVAC systems, lower energy bills, savings on equipment costs, and many other benefits that will be passed on to the home buyer.

The Contract Agreement

Once a good HVAC contractor is identified, a contract agreement should be created that establishes the expectations of both parties, particularly in the area of the scope of work. The scope of work should be adapted from the bid or included in the contract by reference to the bid. A contract is important for many reasons, and there are many good resources for help in this area (see, for example, *Contracts and Liability for Builders and Remodelers*, 4th ed., National Association of Home Builders (NAHB), 1996).

The contract agreement defines a relationship through which quality may be built in. The largest complaint of HVAC contractors is that they do not get paid on time. They are often ill-informed when changes affect their work. They also get called on the job at the wrong time. All of these conditions affect productivity, quality, and ultimately the performance of the HVAC system and ductwork. If these basic issues are not reasonably controlled, then quality HVAC systems and ductwork may be the least of your concerns.

PLANNING

Planning and coordination involve careful consideration of the distribution system during the house design phase as well as accurate scheduling of HVAC installation. The initial stages of house design define the building that the HVAC system will condition but may exclude meaningful planning for the system itself. Since a forced-air distribution system requires significant space and is strongly affected by the framing and floor plan, consideration should be given to its design and layout during this stage.

A simple example of this process is that of a builder who wants to condition the second floor of a two-story house with ceiling registers. The ceiling registers are to be fed from ducts in the attic, while the furnace and system fan are located in the basement. In order to route air from the basement to the attic ducts, a vertical chase will be needed. Planning the location of this chase will minimize its impact on living space and provide a "straight shot" to the attic, instead of a pathway that is too small or has elbows or bends (which would add cost as well). A possible solution would be to frame the chase into the back of interior closets on the first and second floors to provide a straight, properly sized path with minimal effect on living space in the home.

Problems are likely to result if architectural plans are developed with the HVAC system simply represented by equipment location and a line drawing for ductwork. With this lack of planning, ducts may be squeezed in where space allows at the time of installation. Any of these conditions will result in reduced system performance, lower occupant comfort levels, and a loss of energy efficiency.

COORDINATION

A builder's coordination of the HVAC system design and installation is a key activity. By overseeing the entire HVAC process with a quality-minded approach, your coordination serves to:

- Develop thorough initial plans.
- Achieve an appropriate design.
- Provide accurate scheduling.
- Identify quality solutions to onsite problems.
- Prevent recurrence of defects.
- Verify and preserve duct system quality after installation.

Planning initially matches the system's specifications and layout to the proposed house, as discussed above. Coordination continues this process in the construction phase. Addressing onsite questions and problems with practical and timely solutions saves time and money, and preserves the quality of the distribution system. For instance, if a plumbing pipe is routed in the same joist bay where a duct takeoff was to be located, the HVAC contractor and site superintendent could choose an appropriate alternative that may be corrected onsite and then red-lined on plans for future sites. The solution may involve moving the ductwork on the current site, but moving the plumbing on future sites—whatever makes sense. This process is much better than trying to force the duct takeoff next to the plumbing pipe. The takeoff duct would have to be reduced inside or routed into the next joist bay, which would restrict airflow to the register. This problem is made worse if left undetected and repeated on several homes.

Jobsite Inspections

Verifying HVAC and duct system installation is another simple task that promotes quality. It can occur at the same time as inspections of other parts of the home. With a little training, a builder's site superintendent can easily perform brief inspections of the HVAC installation at different stages of construction. The purpose of the inspections should be to identify any deviations from the contract agreement and to detect defects that require corrective action. Effective inspections must be scheduled and documented. Sample inspection checklists are provided on the following pages for that purpose. The inspection documentation must also track the final completion of any corrective actions that are required.

Duct Installation Inspection

Name: _____   Date: _____

JOBSITE: HVAC CONTRACTOR:

                                          CONTRACT #:

Schedule Inspection Several Days Prior To HVAC Rough-In

Framing and structure built as planned: Yes ____   No ____

If "No," action: ☐ correct framing
                    ☐ revise duct layout
                    ☐ other

Orientation of house as planned: Yes ____   No ____

If "No," action: ☐ revise room airflows, duct sizes
                    ☐ other

Glazing quantity and locations as planned: Yes ____   No ____

If "No," action: ☐ revise room airflows, duct sizes
                    ☐ other

Comments & Corrective Actions (for "No" responses):

_____

_____

_____

_____

_____

_____

_____

_____

_____

Job-Ready Inspection

Name: _____   Date: _____

JOBSITE: HVAC CONTRACTOR:

                                            CONTRACT #:

Schedule Inspection Just Prior To Completion Of Duct Installation

Trunk and takeoff sizes as planned:	Yes _____	No _____
Each segment supported:	Yes _____	No _____
Segments mechanically fastened together:	Yes _____	No _____
Number of takeoffs as planned:	Yes _____	No _____
Takeoffs securely fastened to trunk:	Yes _____	No _____
Sealing requirements (if any) as planned:	Yes _____	No _____
Duct insulation (where required) meets R-value requirements:	Yes _____	No _____
Duct insulation (where required) covers duct completely (no gaps at joints):	Yes _____	No _____
Dampers (if applicable) installed as planned:	Yes _____	No _____
Correct number of supply and return registers:	Yes _____	No _____
Building cavity return ducts (if applicable) sealed off with blocking material:	Yes _____	No _____

Comments & Corrective Actions (for "No" responses):

_____

_____

_____

_____

Close-In Inspection

Name: _____   Date: _____

JOBSITE: HVAC CONTRACTOR:

                                           CONTRACT #:

Schedule Inspection Just Prior To Drywall Installation

Duct to boot joints are secure: Yes ____   No ____

Boot faces are flush with floor surface or surface of the
wall or ceiling frame: Yes ____   No ____

Wall cavity returns (when used) are sealed and airtight
at all penetrations: Yes ____   No ____

Soffits and chases adjacent to unconditioned spaces
are sealed off from these areas: Yes ____   No ____

Concealed ducts have not been pinched, crushed,
ripped, or disconnected: Yes ____   No ____

Equipment to plenum joints are airtight: Yes ____   No ____

Ducts in exterior wall cavities (when used) have
adequate insulation between duct and exterior: Yes ____   No ____

Comments & Corrective Actions (for "No" responses):

_____

_____

_____

_____

_____

_____

_____

Final Inspection

Name: _____   Date: _____

JOBSITE: HVAC CONTRACTOR:

                                    CONTRACT #:

Schedule Inspection Just Prior To Building Turnover

    Return filter(s) clean and in place: Yes _____   No _____

    Registers securely seated into boots: Yes _____   No _____

    System fan produces flow out of every register: Yes _____   No _____

    Visible ducts are properly connected and not damaged
    or pinched-off: Yes _____   No _____

Comments & Corrective Actions (for "No" responses):

_____

_____

_____

_____

_____

_____

_____

_____

_____

_____

_____

_____

_____

_____

Chapter 6

Marketing Duct Efficiency

As a builder looking to market duct and energy efficiency, you'll need to identify a target market, study available utility energy programs, complete energy analyses for home energy ratings, offer energy financing options, and develop a complete marketing plan. The following sections provide information which will help you complete these tasks.

IDENTIFY THE MARKET

Energy-efficient ductwork, products, and construction practices can be a tough sell. Features can be expensive and confusing and payback periods can be vague. Also, home buyers from different market segments may want different features for different reasons. Some reasons that home buyers may not ask about the energy-efficient features of a home may include:

- Energy-efficient products and construction practices are not visible or tangible the way jacuzzis or crown molding are.
- Energy topics are technical, can be confusing, and can require specialized knowledge.
- Energy efficiency just adds more options to a long list of choices, overwhelming the buyer.
- Energy-efficient products and practices add cost.
- New homes are assumed to be more efficient.

Left with a confusing array of additional costs, a buyer will simply withdraw or reduce discussions to floor plans and interior finishes. You'll need to develop an active marketing strategy that identifies target market segments, defines the features and benefits that appeal to each segment, and develops an effective marketing and sales approach.

Identify Market Segments

As a builder, you must first identify the market segment to which you're going to sell:

- First-time home buyer.
- Move-up home buyer.
- Custom home buyer.
- Luxury home buyer.

Buyers from each segment have different needs and respond to different benefits and features. For instance, first-time home buyers might be expected to know little about energy efficiency. Not only are they busy trying to come up with the financing to get into their first home, they do not have the home owner-ship experience or expectations of move-up buyers. First-time home buyers, however, may have the most to gain from energy efficiency features which qualify them for a higher loan through an energy-efficient mortgage (EEM).

Move-up home buyers have experience to rely on. For instance, they may have experienced poor comfort in a previous home and, as a result, would place high value on good HVAC design and installation. They may also be sophisticated enough to decide that the price of efficiency improvements that yield lower monthly utility costs makes sense when bundled into an interest-deductible mortgage. Custom home buyers will have in mind specific features for their home, based on specific needs. By identifying the basic needs, additional features may be offered as options. Luxury home buyers are often investing more money in a larger home and expect to have optimal comfort or lower utility bills. Efficient ductwork, equipment, and construction features are attractive options for large homes with huge monthly utility bills.

Convert Features to Benefits

Features are the physical characteristics of a home, such as the types of materials and specific products or construction techniques used to build it. Features are used to differentiate competitors. Each feature has performance characteristics or advantages that may attract different buyers. Benefit statements communicate the value of those advantages to buyers.

Typical energy-efficient features include:

- **HVAC Equipment.** High efficiency furnaces, air conditioners, and heat pumps; zone dampers, air cleaners, humidifiers, and setback thermostats.
- **Ductwork.** Improved duct location, duct insulation, and duct sealing.
- **Wall and Ceiling Insulation.** Increased thickness and density of insulation, reduced lumber framing, high heel roof trusses, and so forth.
- **Windows.** Low-e coatings, gas filled units, more efficient spacers and frame materials.
- **Air Tightness Detailing.** Sealing drywall or sheathing, caulking band joists, sealing window and door openings and electrical/plumbing penetrations.
- **Water Heating Equipment.** Higher efficiency water heaters.
- **Lighting and Appliances.** Fluorescent lighting, efficient refrigerators, washing machines, and so forth.

All energy features will offer some mix of efficiency benefits. Customers will vary in their response to different benefits. Some of the benefits of increased efficiency include:

- **Increased Comfort.** Better, more energy-efficient construction can provide more uniform temperatures, better humidity control, and less draftiness.
- **Lower Utility Bills.** An obvious goal of energy efficiency. Mortgage increases for improved energy-efficient products and construction practices are tax-deductible. Utility bills are paid with after-tax income.
- **Financing Options.** Special mortgage programs are available for energy-efficient homes (see Financing chapter).
- **Improved Health and Indoor Air Quality.** Tight construction controls for the entry of radon/soil gases, pollens, spores and other outdoor pollutants. Indoor pollutants can be controlled through design and point ventilation.
- **Increased Resale Value.** While little hard data exists, added value at the time of sale should translate to added value at resale.
- **Better Perceived Quality.** The design and detailing requirements of energy efficiency lead home buyers to associate energy efficiency with quality.
- **Lower Atmospheric Pollution.** Energy efficiency is generally associated with reduced carbon dioxide gases and other emissions from coal and gas combustion.
- **State-of-the-art Technology.** The technology buff might be interested in how the energy efficiency features function, and may simply want to own the latest in technology.

Marketing Process: Getting the Word Out

Once you've defined a market and the features valued by that market, you need to communicate the benefits of these features to home buyers. Marketing energy efficiency requires consistent messages communicated in multiple ways. An effective communication strategy includes advertising, onsite marketing, and knowledgeable sales staff. If energy benefits are communicated effectively, a potential customer will remember what makes an energy-efficient home different from a competing home.

Advertising

Ideas for consistently advertising energy efficiency in print, direct mail, and broadcasting media include the following:

- Program name and logos that customers identify with energy-efficient homes.
- Comparative graphs that illustrate the energy efficiency of competitors' homes.
- Testimonials on low operating costs and comfort from satisfied customers.
- Guarantees that a home's operating costs will not exceed a certain amount. This may sound impossible because of all of the different variables that need to be considered, but builders around the country are using this selling tool. It is possible if you know a home's infiltration and duct leakage.
- Ratings that allow customers to easily compare the energy efficiency of homes like miles per gallon on a car.

- Financing opportunities which allow customers to qualify for more home or to add the cost of energy upgrades into the mortgage, which makes the savings appear as after-tax income.
- Technical information, including specifications and efficiencies which allow interested home buyers to understand and compare different options.
- Tests and certification of the home which establish and verify its compliance to specific standards and overall energy efficiency.

Onsite Marketing

Onsite marketing reinforces advertising messages. Use your program name and logo consistently in all onsite marketing materials. Successful builders in the energy niche constantly reinforce the image that they are different. This type of marketing can include:

- Site signs which are consistent with advertising messages so that customers know they have arrived at the right place.
- Cut-away wall sections in model homes which show how the walls are constructed to add efficiency.
- Diagrams, posters, magnetic signs, and samples of heating and cooling equipment, sealed duct sections, and other products can be used to demonstrate technologies.
- Energy features, specifications, and warranties with blank spaces provide home buyers with a checklist to shop and compare when they visit competitors.
- Framed testimonials from satisfied customers can be viewed while customers are onsite.

Give home buyers technical information, but make the messages simple, clear, and in terms they understand (e.g., "year-round comfort and guaranteed energy bills"). Make sure sales staff and real estate agents know that this information is available and keep it visible to customers.

Sales Technique

Nothing is more important than a trained and motivated sales staff who can answer questions and provide information that the potential buyer may not have considered. Here are a few ideas to try with your sales staff:

- Provide onsite training of sales staff and real estate agents with construction superintendents, so they really understand energy efficiency features and benefits.
- Offer offsite training for sales people with groups like NAHB and the National Association of Realtors®. Both organizations have training programs for sales people and agents which cover new home construction and, to some degree, energy efficiency.
- Provide sales staff and real estate agents with incentives to sell energy efficiency upgrades.

- Remind potential customers of energy efficiency features and benefits in follow-up sales calls.

A key objective of the sales process should be to match the features and benefits of the home to the specific needs or interests of the buyer. The following table provides suggestions for identifying energy-related benefits that would appeal to your customer.

If the customer . . .	Discuss this benefit . . .
• Complains about comfort in current home • Asks about type of heating system • Mentions they moved from New England to avoid cold winters	Comfort
• Complains or comments on costs of energy or maintenance of current home • Asks about maintenance schedule or cost of any component of home • Asks about the specific price of many upgrades	Lower Utility Bills
• Expresses concern about ability to get loan • Asks if the home is eligible for any rebates or special financing programs	Financing Options
• Asks about out-gassing from building materials (e.g., formaldehyde) • Mentions "sick buildings" • Mentions that family members have asthma or allergies • Mentions being a non-smoker • Mentions membership or interest in environmental organizations • Asks about CFCs in building materials • Asks about recycling	Indoor Air Quality and the Environment
• Mentions making a profit or loss on resale of existing home • Asks about resale history in the neighborhood	Increased Resale Value
• Discusses quality or construction integrity of current home, or claims older homes are better built • Comments on any features of the home as an indicator of quality or integrity	Better Quality
• Asks for definitions of terms like "Low-e" or "SEER" • Mentions training as an engineer or scientist • Asks if blower door or other performance tests have been done on the home	Technology

ENERGY EFFICIENCY PROGRAMS

Energy Analyses

Two basic types of energy calculations are commonly used in residential design. The first is *design load* calculations, in which you estimate the maximum heating load and maximum cooling load, for purposes of selecting the proper size of heating and cooling system. These calculations are performed for "design conditions," which are approximately the most extreme conditions expected during the year. The results are a rate of heating or cooling required in Btu per hour to provide comfort under the design conditions. These calculations have traditionally been performed by HVAC contractors or distributors.

The other category of energy calculations is *annual energy use* estimations. These are calculations of the expected amount of energy that will be used in the home during a typical year for heating, cooling, water heating, lights, and appliances. Estimation methods may be performed by hand, such as the classic "degree day method" for estimating heating energy, or by using packaged computer software. Computerized estimations use more sophisticated calculation methods.

Presentation of energy use estimations can be the core of a strategy for marketing energy efficiency. Energy use estimation provides customers with quantitative values that can be compared to their own experience or to other homes for which similar estimations are available.

Utility Programs

Utilities in many locations offer energy efficiency programs for new homes. Most such programs are operated by electric utility conservation and load management or marketing departments. Some of these programs are designed locally, and some are provided by program operators such as Good Cents or Comfort Home. Many of these programs provide:

- Builder training.
- Certification of an individual home as meeting program requirements.
- Energy-performance testing as part of the certification process.
- Marketing materials.
- Financial incentives to achieve participation.

Though fewer in number, similar programs are available through some gas utilities.

There are three basic approaches to compliance used in utility programs: prescriptive compliance, alternate paths, and performance verification.

- Prescriptive compliance programs require builders to include specific energy features, (e.g., R-15 walls, R-30 ceilings, R-6 duct wraps, and so forth). This approach is easily understood and enables straightforward compliance, but offers no design flexibility.

- Alternate paths offer the option of several prescriptive packages, e.g., one might use either R-15 walls with low-e argon glass, or R-19 walls with standard double-pane windows. This provides some design flexibility to the builder.
- Performance verification programs establish a maximum energy consumption requirement. This allows builders great flexibility, though compliance verification is generally more complex.

Incentive structures and amounts vary among utilities. Some provide incentives for the builder to incorporate energy upgrades. These can range up to $2,000 or even more per home to reimburse builders for upgrades required to meet a compliance standard. Other incentives are given directly to the homeowner to cover first costs associated with purchasing a more expensive, energy-efficient home. One interesting approach offered by an electric cooperative in southern Maryland is to give first-time home buyers the incentive, to be used as part of their down payment. This is done in partnership with a local lender under the Fannie Mae Community Home Buyer's Program.

Marketing materials provided by these programs often include logos, brochures on features and benefits, and local advertising. In general, the value of community-wide name recognition associated with utility programs is a compelling reason for participation in such programs, rather than creating a program independently. By providing third-party confirmation that a consumer is buying an energy-efficient home, these programs also add a perception of credibility and quality assurance.

Builders interested in utility energy program opportunities should call a customer service representative at their local utility.

Home Energy Ratings

Over the years, many observers have commented that it would be useful to have a standard system for reporting the relative energy efficiency of homes. The use of miles per gallon ratings for automobiles is often mentioned as an analogy. Such systems have been developed and are called Home Energy Rating Systems (HERS). A collaborative group representing builders, utilities, government agencies, and other interested parties has formulated standards for HERS which would make all ratings more or less equivalent. While the proposed national standard to create a uniform rating methodology has not been adopted as of this writing, most providers of ratings are moving to the proposed standard.

Under the proposed standards for home energy ratings (and using one of several software-based methods), the energy use of a home in any location is compared to a home of the same size and shape but having energy features that just meet the CABO Model Energy Code (MEC) requirements for the area. The rating is based on the relative energy use of the home as designed to the home with MEC features. Under this system, ratings are reported in three ways: as a number of stars from "One Star" to "Five Star Plus," as a relative value from one to 100, and by estimated total annual cost. The more stars or higher numerical rating, the more efficient the home.

The results of HERS and other energy analyses can be used to communicate several kinds of messages about energy efficiency:

- Certificates and stickers call attention to the energy efficiency of the home.
- Comparative ratings can show the advantages of a home over older existing housing, over other new homes that do not incorporate energy efficiency features, and even over other energy-efficient homes that use different features.
- Comparative ratings can show the differences among available energy option packages.
- Analyses can be the basis for energy cost guarantees. (Note that most ratings do not provide a guarantee of energy use.)
- Rating certificates provide a third-party validation of energy efficiency.

HERS and other analysis results can be presented in a variety of ways, including stickers placed in a permanent location in the home, certificates displayed in the home or given to the owner at time of sale, and graphics showing predicted energy use on display in the model home.

In general, energy use analyses and home energy ratings are provided either by utilities as part of energy programs for new homes (sometimes at low or no cost), or by private consultants on a for-profit basis. Some states are operating HERS programs that are available to builders and consumers. HVAC contractors are also potential providers. An HVAC contractor or distributor who is performing an analysis to size heating and cooling systems may have the capability to perform an annual energy use prediction with little extra effort.

The HERS council may be contacted at:

Home Energy Rating System Council
1511 K Street, NW, Suite 600
Washington, DC 20005
(202) 638-3700

ENERGY EFFICIENCY FINANCING

Energy efficiency financing is any mortgage or program that provides financing incentives for the purchase of an energy-efficient home, or for the purchase or refinancing of an existing home. Such financing may be offered by local lenders, the secondary market (e.g., Fannie Mae and Freddie Mac), and utilities. Mortgage markets change quickly and are very dependent on what local lenders and states are willing and able to provide. Some general financing opportunities are discussed below.

Energy-Efficient Mortgage

An Energy-Efficient Mortgage, or EEM, also known as the "stretch mortgage" allows the purchasers of an energy-efficient home a two-percent stretch on their debt-to-income and loan-to-income qualifying ratios. The debt-to-income ratio, which is usually 28 percent, includes all monthly expenses. The overall debt ratio is typically 36 percent and includes the total housing expenses, including loan principle, interest, taxes and insurance (PITI). By adding a 2 percent stretch to both ratios in addition to all other compensating factors, the potential home-owner qualifies for a more expensive home with lower income.

Criticisms of this type of EEM are that:

- People are not interested in higher qualifying ratios unless interest rates are high.
- The appraisal may not justify the added borrowing.
- There is too much paperwork to justify the stretch and have it accepted by the secondary market.

Energy-Improvement Mortgage

Another general type of energy efficiency financing is the energy-improvement mortgage. In this type of financing, the mortgage amount is increased by the cost of qualifying energy improvements. The energy upgrade costs included must prove "cost-effective," meaning that the present value of the improvements must be calculated and demonstrate that monthly dollar savings exceed the additional mortgage payment.

Other Financing Options

Other options include reduced down payments and reduced interest rate financing. Although less common, local lenders and utilities in some areas have teamed together to provide reduced down payments to low- and moderate-income homeowners. In one case, utility incentives for energy-efficient homes goes directly to the home buyer as part of the down payment. The Fannie Mae Community Home Buyers Program requires a potential homeowner to pay only 3 percent of a 5 percent down payment, with the other 2 percent provided as a gift. Also, in a few cases, private lenders and state programs have provided reduced interest rate financing for energy-efficient homes.

To learn more about energy efficiency financing opportunities, or to find a lender near you, contact the Residential Energy Services Network (RESNET), a joint project of Energy Rated Homes of America and the National Association of State Energy Officials. RESNET can be reached at: (907) 345-1930 or http://www.natresnet.org

PUTTING IT ALL TOGETHER

The following guidelines are offered as an aid in marketing energy efficiency.

What is Your Energy Program?

- Define your personal approach to energy efficiency. If you participate in a utility program that allows multiple paths to compliance, what path are you choosing, and why? Have you considered the visibility and salability of the methods you choose? For example, you may decide that the efficiency label on an air conditioner makes its efficiency more visible and marketable than insulation hidden in the wall or, conversely, that insulation is intuitively more understandable to sales people and buyers than high efficiency air conditioning.

- If you do not participate in a utility program, you have a much larger job. You need to have an understandable program, including specific energy efficiency features that are standard or available in one or more option package. Even in custom building starting with a standard package will help establish your specific identity as a builder of energy-efficient homes. You will want to establish specific procedures for validation and quality assurance of your energy systems.

- Finally, you will want to establish a marketing identity for your program. Naming your program may help. Consistent use of a logo from one or more manufacturers of energy-related materials or components can combine name recognition to the energy message. Legitimate statements of energy savings or energy use that characterize your homes will be of value.

Constant Use of Name, Features, and Benefits

- The utility program name or your own program name and logo should appear on every advertisement and all printed material you produce. Don't lose a potential customer by doing this halfway. The most successful builders offering energy efficiency are careful to get their name and logo in front of customers at every chance.

- Identify the features you use in your houses to make them more efficient. Some buyers may not understand what they are, or may not care. Others, however, will ask questions about them, will ask competing builders about their energy features, and may want to discuss energy in more detail. Listing features invites this discussion, and it will benefit you.

- Whenever discussing energy efficiency features, mention at least the most basic benefits. Utility costs and comfort are the benefits most attractive to buyers. Remember that energy is technical, and don't assume buyers will automatically make the jump from features to benefits.

Advertising and Marketing Ideas

There are dozens of advertising and marketing techniques for communicating your message. Here are a few you might want to consider:

- Energy program certifications.
- Manufacturers' brochures on energy options.

- Energy analysis or HERS results.
- Testimonials from satisfied customers.
- Cutaway displays of wall construction, and so forth in model homes.
- Photos of construction showing hidden energy features.
- Videos of energy efficiency in process, including blower door testing, and so forth.
- Blower door testing performed with customer onsite.

Knowledgeable Sales Staff

- A motivated and knowledgeable sales staff, either in-house or a real estate agent, is critical to success in selling energy efficiency. The first step to good sales is achieving a basic level of technical understanding. Methods for training sales people include utility programs, in-house training with or by construction superintendents, trade journals, and construction- and energy-related educational programs offered by NAHB and the National Association of Realtors®.
- Sales staff should understand all the specifics of your program, including standard and optional energy features, package costs, brand names and specifications, utility program requirements and incentives, and any financing packages available.
- Finally, sales people should spend time with customers owning energy-efficient homes, to get the story directly and thus be able to pass it on more effectively. Using these techniques will make increased energy efficiency a winner for you, your sales staff, and your customers.

References

SYSTEM COMPONENTS

1. *HVAC Duct Construction Standards: Metal and Flexible*, Sheet Metal and Air Conditioning Contractors National Association, Inc. (SMACNA), 1st ed., 1985.

2. *Fibrous Glass Residential Duct Construction Standards*, Sheet Metal and Air Conditioning Contractors National Association, Inc. (SMACNA), 6th ed., 1992.

3. *Installation Standards For Residential Heating and Air Conditioning Systems*, Sheet Metal and Air Conditioning Contractors National Association, Inc. (SMACNA), 6th ed., 1988.

4. *Duct Board Fabrication Concepts*, Instruction Video Tapes, North American Insulation Manufacturers Association (NAIMA), 1994.

5. *A Guide To Insulated Air Duct Systems*, North American Insulation Manufacturers Association (NAIMA), 1993.

6. *Fibrous Glass Duct Construction Standards*, North American Insulation Manufacturers Association (NAIMA), 1993.

7. *Fibrous Glass Duct Construction With 1-1/2 Inch Duct Board*, North American Insulation Manufacturers Association (NAIMA), 1993.

8. *Fibrous Glass Residential Duct Construction Standards*, North American Insulation Manufacturers Association (NAIMA), 1993.

9. *Flexible Duct Performance and Installation Standards*, Air Diffusion Council (ADC), 1991.

SYSTEM SELECTION

1. *Improved Air Distribution Systems For Forced-Air Heating*, RCDP IV Final Report, Bonneville Power Administration, July 1995.

2. *Manual D: Residential Duct System*, Air Conditioning Contractors of America, 1995.

3. *Manual T: Air Distribution Basics for Residential and Small Commercial Buildings*, Air Conditioning Contractors of America.

4. *Manual G: Selection of Distribution System*, 1st ed., Air Conditioning Contractors of America.

SYSTEM DESIGN

1. *Heat Pump Manual*, EPRI EM-4110-SR, Electric Power and Research Institute (EPRI) and National Rural Electric Cooperative Association (NRECA), 1985.

2. *Manual H: Heat Pump Systems, Principles and Application*, 2nd ed., Air Conditioning Contractors of America, 1984.

3. *Manual S: Residential Equipment Selection*, 1st ed., Air Conditioning Contractors of America.

ENERGY-EFFICIENT PRACTICES

1. ANSI/ASHRAE 103: Method and Testing For Annual Utilization Efficiency of Residential Control Furnaces and Boilers, 1993.

2. *Manual J: Load Calculation For Residential Winter and Summer Air Conditioning*, Air Conditioning Contractors of America, 7th ed., 1986.

3. *Manual D: Residential Duct Systems*, Air Conditioning Contractors of America, 1995.

4. *ASHRAE Handbook: HVAC Systems and Equipment*, I-P edition, American Society of Heating, Refrigerating, and Air Conditioning Engineers, Inc., 1993.

5. *Heat Pump Manual*, EPRI EM-4110-SR, Electric Power and Research Institute (EPRI) and National Rural Electric Cooperative Association (NRECA), 1985.

6. CABO One and Two Family Dwelling Code, BOCA, SBCCI, and ICBO, 1995.

7. Model Energy Code, 1995.

MARKETING DUCT EFFICIENCY

1. *What Today's Home Buyers Want*, National Association of Home Builders (NAHB), 1996.

2. *Residential Energy Efficiency Program*, Interim Reports, NAHB Research Center, 1996.

Additional Resources

These publications and additional resources on construction topics are available from the NAHB Research Center or NAHB's Home Builder Bookstore.

CABO Model Energy Codes (1992 or 1993)
NAHB Home Builder Bookstore

Contracts and Liability for Builders and Remodelers (book and diskette)
NAHB Home Builder Bookstore

Design/Frost-Protected Shallow Foundations
NAHB Research Center or Home Builder Bookstore

Energy-Smart Building for Increased Quality, Comfort, and Sales
NAHB Home Builder Bookstore

Model Energy Code Thermal Envelope Compliance Guide
NAHB Home Builder Bookstore

NAHB Building Code Action Kit
NAHB Home Builder Bookstore

Production Checklist (book and diskette)
NAHB Home Builder Bookstore

To request a catalog or obtain more information about a particular publication, contact:

NAHB Research Center, Inc.
400 Prince Georges Boulevard
Upper Marlboro, MD 20772-8731
(800) 638-8556
http://www.nahbrc.com

Home Builder Bookstore
National Association of Home Builders
1201 15th Street, NW
Washington, DC 20005-2800
(800) 223-2665
http://www.nahb.com/builderbooks